D0206157

Physiology and Ecology of Fish Migration

Physiology and Ecology of Fish Migration

Editors

Hiroshi Ueda
Laboratory of Aquatic Bioresources and Environment
Field Science Center for Northern Biosphere
Division of Biosphere Science
Graduate School of Environmental Science
Hokkaido University
Sapporo
Japan

Katsumi Tsukamoto
Department of Marine Bioscience
Atmosphere and Ocean Research Institute
The University of Tokyo
Kashiwa, Chiba
Japan

CRC Press
Taylor & Francis Group
Boca Raton London New York

CRC Press is an imprint of the
Taylor & Francis Group, an **informa** business

A SCIENCE PUBLISHERS BOOK

CRC Press
Taylor & Francis Group
6000 Broken Sound Parkway NW, Suite 300
Boca Raton, FL 33487-2742

Cover Illustration: Reproduced by kind courtesy of Teruhiko Awakura

Library of Congress Cataloging-in-Publication Data

Physiology and ecology of fish migration / editors, Hiroshi Ueda, Katsumi Tsukamoto.
        pages cm
  Includes bibliographical references and index.
  ISBN 978-1-4665-9513-2 (hardcover : alk. paper) 1. Fishes--Migration. I. Ueda, Hiroshi, 1951- II. Tsukamoto, Katsumi.

  QL639.5.P49 2013
  597.1568--dc23
                                        2013010317

Visit the Taylor & Francis Web site at
http://www.taylorandfrancis.com

CRC Press Web site at
http://www.crcpress.com

Science Publishers Web site at
http://www.scipub.net

Preface

Among the roughly 30,000 species of fish, migratory species account for only 165 species, but they show very interesting biological phenomena and life histories. Most migratory fishes are very important fisheries resources. Migratory fishes move either alone or in schools from their birth place to their growth habitats, and then usually return to their place of birth for reproduction. These migrations allow fishes not only to exploit different food availabilities and to adapt to environmental changes, but also to regulate population density and widen their distribution.

In general, fish migrations are classified into three categories; "Recruitment" migrations from juvenile nursery area to adult habitat, "Contranatant" and "Denatant" migrations between adult habitat and adult spawning areas, and "Drift" or "Denatant" migrations from adult spawning area to juvenile nursery area (Harden-Jones 1968) (Fig. 1). These migrations form a migration loop (Tsukamoto et al. 2002). There must be very important initiation (triggering) and adaptation mechanisms that make it possible for fish to migrate from juvenile nursery area to adult habitat and from adult habitat to adult spawning area, but many of these mechanisms remain a mystery.

The recent rapid development in biotelemetry techniques, such as ultrasonic and radio telemetry, data logging and pop-up satellite telemetry, make it possible to investigate wild fish behavior both in freshwater and seawater. Following on from the Conference on Fish Telemetry organized in Europe that had been held in Belgium in 1995, France in 1997, United Kingdom in 1999, Norway in 2001, Italy in 2003, Portugal in 2005, Denmark in 2007, and Sweden in 2009, the memorable 1st International Conference on Fish Telemetry (ICFT) was held in Japan in 2011. Many authors of this book, *Physiology and Ecology of Fish Migration*, presented their most updated, innovative researches in the 1st ICFT. This book is a part of the conference proceedings.

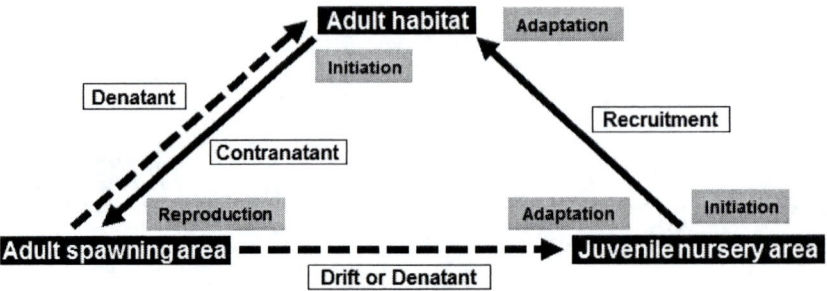

Fig. 1. Migration loop of fish. Revised form Harden-Jones 1968.

Each chapter deals with the following topics:

Chapter 1: Physiology of imprinting and homing migration in Pacific salmon

Chapter 2: The behavior and physiology of migrating Atlantic salmon

Chapter 3: The onset mechanism of the spawning migrations of anguillid eels

Chapter 4: Marine migratory behavior of the European silver eel

Chapter 5: Sea lamprey migration: a millenial journey

Chapter 6: Migratory behavior of adult Pacific lamprey and evidence for effects of individual temperament on migration rate

Chapter 7: Behavioral ecology and thermal physiology of immature Pacific bluefin tuna.

We sincerely hope that this book will contribute to promote a better understanding of the physiology and ecology of fish migration.

Hiroshi Ueda
Katsumi Tsukamoto
Sapporo and Tokyo, Japan.

References

Harden-Jones, F.R. 1968. Fish Migration. Arnold Press, London. UK.
Tsukamoto, K., J. Aoyama, and M.J. Miller. 2002. Migration, speciation, and the evolution of diadromy in anguillid eels. Can. J. Fish. Aquat. Sci. 59: 1989–1998.

Contents

CHAPTER 1

Physiology of Imprinting and Homing Migration in Pacific Salmon

Hiroshi Ueda

Introduction

Salmon have an amazing ability to migrate thousands of kilometers from the ocean to their natal stream for reproduction. The reproductive homing migration is one of the most interesting mysteries of the salmon life cycle and most challenging to study. It is now believed that some specific factors of the natal stream are imprinted in particular nervous systems of juvenile salmon during downstream imprinting migration, and that adult salmon evoke these factors to recognize their natal stream during upstream homing migration (Ueda 2012). Since the olfactory hypothesis for salmon imprinting and homing to their natal stream was proposed by Hasler's research group in the 1950s (Hasler and Scholz 1983), this olfactory discriminating ability is believed to be exerted within a short distance from the coast of their natal stream. It is also impossible for salmon to use this ability only for long distance migration from the feeding area to their natal stream. It is therefore still unknown which sensory systems play leading roles in open water orientation, which endocrine hormones control imprinting and homing migration, and how the olfactory system discriminates between various stream odors.

Laboratory of Aquatic Bioresources and Environment, Field Science Center for Northern Biosphere, Division of Biosphere Science, Graduate School of Environmental Science, Hokkaido University, North 9 West 9, Kita-ku, Sapporo, 060-0809, Japan.
Email: hueda@fsc.hokudai.ac.jp

In this review, the following research topics are described to clarify the physiology of imprinting and homing migration in Pacific salmon—life cycle comparison among four Pacific salmon in Japan, physiological biotelemetry researches on salmon homing migration, endocrinological researches on salmon imprinting and homing migration, and neurophysiological researches on olfactory imprinting and homing abilities. These topics discuss the evolutional aspects of four Pacific salmon, navigation abilities during homing migration in open water, hormonal controlling mechanisms of imprinting and homing migration, and olfactory imprinting and discriminating capabilities of natal stream odors.

1. Life Cycle Comparison Among Four Pacific Salmon in Japan

There are four Pacific salmon (genus *Oncorhynchus*) species in Japan: pink salmon (*O. gorbuscha*), chum salmon (*O. keta*), sockeye salmon (*O. nerka*), and masu salmon (*O. masou*). The life cycles of the former two species are quite different from the latter two species (Fig. 1). All juvenile fry of pink and chum salmon migrate downstream a few months after emergence, and adult fish migrate upstream a few weeks before final gonadal maturation.

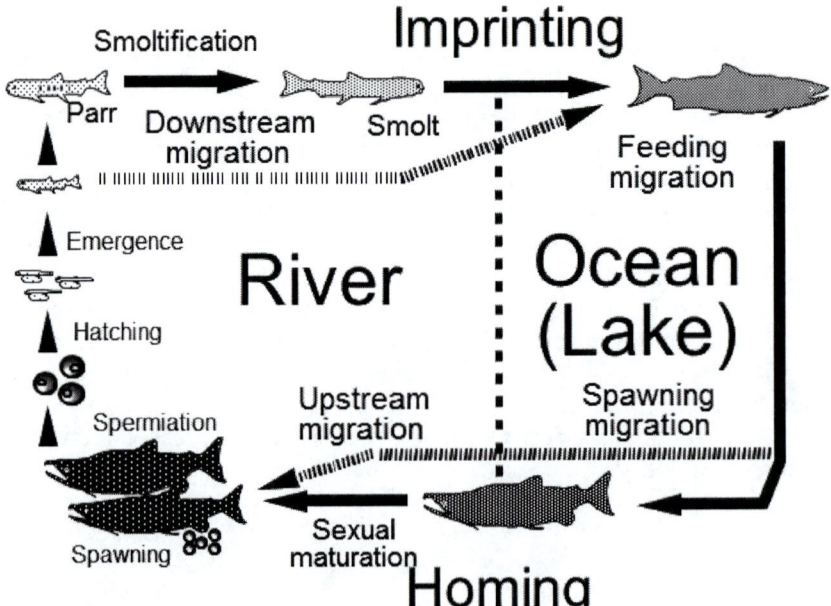

Fig. 1. Life history of two different types of Pacific salmonid species in Japan. Dotted line: chum and pink salmon; Solid line: sockeye and masu salmon.

In contrast, juveniles of the sockeye and masu salmon remain for 16–18 months in streams or lakes to grow into smolts that are able to adapt to seawater, and then conduct downstream migration. Their adults perform upstream migration 4–5 months prior to final gonadal maturation (Groot and Margolis 1991). There are also land-locked forms of both sockeye and masu salmon that remain in freshwater throughout their lives.

In a phylogenetic division of these four species using retropositional genome analyses, pink salmon are considered to be the most advanced while masu salmon are considered to be the most primitive species (Murata et al. 1996). Analyses of the relationship between the oceanic distribution and the population size among these four species reveal that pink salmon are distributed the most widely and have the largest population, while masu salmon appear to have the most restricted distribution and have the smallest population (Kaeriyama and Ueda 1998). Although the homing accuracy of these salmon has not been examined in detail, it is believed that masu salmon come back to their natal stream with the highest percentage, and that pink salmon stray off into the non-natal stream. If salmon always show a highly accurate homing to their natal stream, there would be little chance for them to widen their distribution area or increase their population size; also, there is the dangerous possibility that their genetic diversity would be reduced. The relationship between salmon evolution and homing accuracy is one of the most interesting problems from the point of view of biological evolution.

2. Physiological Biotelemetry Researches on Salmon Homing Migration

The recent rapid development in biotelemetry techniques, such as ultrasonic and radio telemetry, data logging, and pop-up satellite telemetry makes feasible the continuous observation of salmon underwater behavior in open water that was impossible to monitor by previous techniques (Cooke et al. 2004; Ueda 2004). In particular, ultrasonic transmitters that emit pulsed signals have been used to investigate the migratory behavior of salmon in the coastal vicinity of their natal stream (Quinn and Groot 1984; Quinn et al. 1989) and the central Bering Sea (Ogura and Ishida 1994). Moreover, ultrasonic tracking in combination with sensory ablation experiments, which blocked visual and olfactory cues or magnetic senses, have been performed several times with oceanic migratory salmonids (Døving et al. 1985; Yano and Nakamura 1992; Hansen et al. 1993; Yano et al. 1996).

2.1. Chum salmon from the Bering Sea to Hokkaido, Japan

Chum salmon were caught by a longline in June 2000 in the central Bering Sea (56°30'N, 179°00'E) in a healthy condition and were estimated to be of Japanese origin by scale analysis because most Japanese chum salmon juveniles were reared in the hatchery and the width of their scale ring during fry stage was wider than wild salmon from other countries. A propeller data logger, which recorded swimming speed (5 sec sampling), depth (5 sec sampling), and temperature (1 min sampling), was attached externally to the dorsal musculature of the fish anterior to the dorsal fin (Tanaka et al. 2005). We released 27 chum salmon with these data loggers, and retrieved one data logger in September 2000 from a set net on the east coast of Hokkaido, Japan (43°20'N, 145°46'E). This, the first record of swimming profiles of homing chum salmon in the oceanic phase provided data for 67 days over a straight distance of 2,750 km and revealed that average swimming speed, depth, and temperature were 62 ± 12 cm/sec, 10.4 ± 14.7 m, and 9.2 ± 0.2°C, respectively (Fig. 2A). Both swimming speed and depth had two peaks around the time of sunrise and sunset with a further small peak around midnight. The fish showed sequential up-and-down movement near the thermocline during twilight and daytime. These diurnal patterns suggest that homing chum salmon allocate time to foraging and that their strategies differ between day and night (Fig. 2B). These results indicate that homing chum salmon have an ability to navigate in the direction of their home and that transport by water currents may help in successful migration. During migration, salmon must recognize their exact location (map) and compass direction (orientation), and must have a biological clock for accurate homing in open water.

2.2. Lacustrine sockeye and masu salmon in Lake Toya, Hokkaido, Japan

For sea-run anadromous populations, it is difficult to carry out controlled and manipulated physiological experiments as fish move from the sea in their pre-maturation phase to their natal stream where they become mature. In contrast, lacustrine populations offer a good model system for studying homing behaviors from open water to their natal area. Lake Toya (surface area 71 km², average and maximum depth 116 m and 179 m, respectively) is a large caldera lake in Hokkaido, Japan. The homing behaviors of mature lacustrine sockeye salmon, whose sensory cues were impaired, were tracked from the center of the lake to the natal area using the ultrasonic tracking system (Ueda et al. 1998). Both a mature male sockeye salmon equipped with a control brass ring (Fig. 3A-1) and a similar mature male sockeye

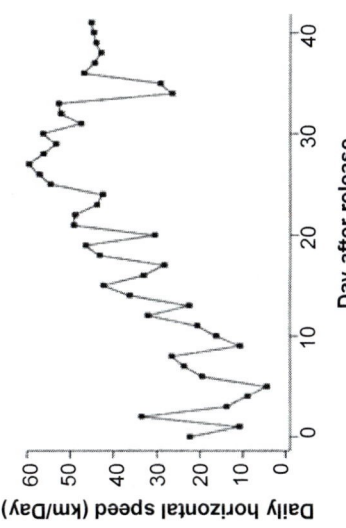

Fig. 2. (A) Swimming depth, ambient temperature, and swimming speed of a Japanese chum salmon from the Bering Sea to Hokkaido, Japan recorded by a propeller data logger. (B) Daily horizontal speed traveled by a Japanese chum salmon from the Bering Sea to Hokkaido, Japan.

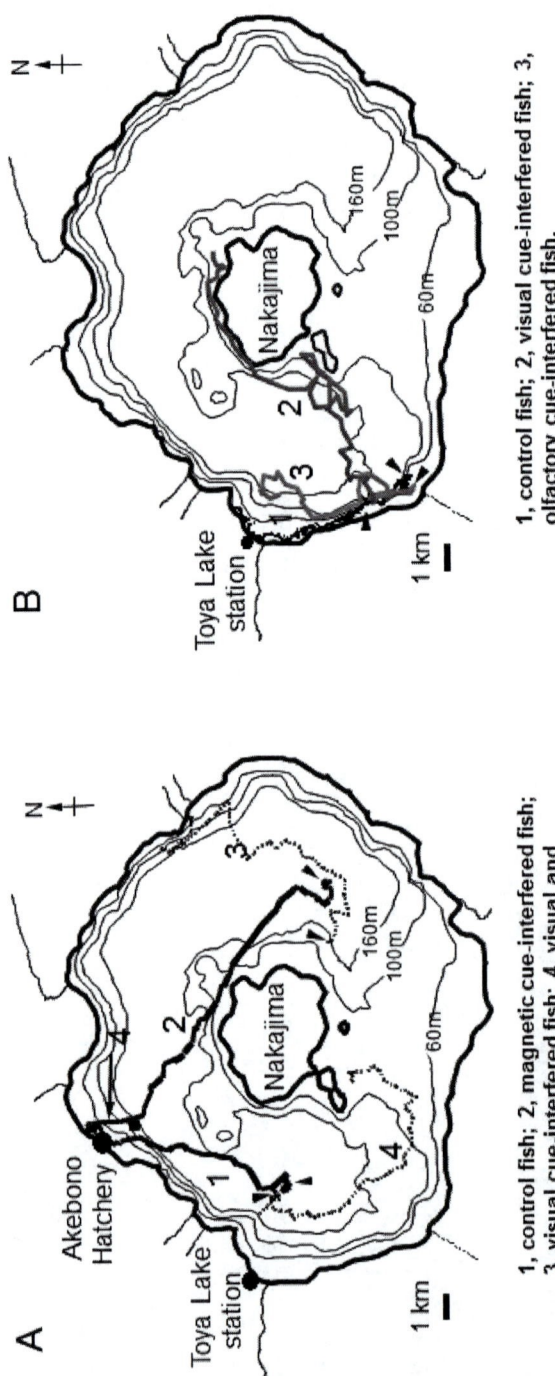

1, control fish; 2, magnetic cue-interfered fish; 3, visual cue-interfered fish; 4, visual and magnetic cues-interfered fish.

1, control fish; 2, visual cue-interfered fish; 3, olfactory cue-interfered fish.

Fig. 3. Tracks of four mature male lacustrine sockeye salmon (A) and three mature lacustrine masu salmon (B) in Lake Toya during the spawning season. Arrowhead indicates the releasing point of each fish.

salmon whose magnetic cues were interfered with by means of a magnetic ring (Fig. 3A-2) returned straight to the natal area after one hour of random movement. A mature male sockeye salmon whose visual and magnetic cues were both blocked moved in a direction opposite to the natal area, but was rediscovered in the natal area on the following evening, suggesting the possible involvement of olfactory cues in finding the natal area (Fig. 3A-3). A blinded male sockeye salmon was also moved to the shore of Naka-Toya far from the natal area in the evening, where it stayed for a few days (Fig. 3A-4). These data suggest that visual cues are critical to the straight homing of sockeye salmon, while magnetic cues do not appear to be necessary for successful return to the natal area. However, magnetoreceptor cells have been identified in the nose of rainbow trout (*O. mykiss*) (Walker et al. 1997). Further study is needed to investigate the involvement of magnetic cues during oceanic homing migration in salmon.

The homing behaviors of mature lacustrine masu salmon were also tracked in Lake Toya (Ueda et al. 2000). A mature control male masu salmon moved constantly along the coast, and stopped his movement at the mouth of a stream (Fig. 3B-1). A blinded mature female masu salmon was released and moved randomly away from the coast (Fig. 3B-2). A mature male masu salmon whose olfactory cue was blocked moved randomly along the coast, and then tended to move away from the coast (Fig. 3B-3). It is interesting to compare the straight movements of sockeye salmon with the coastal movement behaviors of masu salmon. These two species show large differences in oceanic distribution. Sockeye salmon are distributed widely in the North Pacific Ocean, while masu salmon have a more restricted distribution in the western North Pacific Ocean (Kaeriyama and Ueda 1998). These data suggest some ecological aspects of successful homing migration of salmonids, where the more narrowly distributed masu salmon only need coastal recognition ability, but the more widely distributed sockeye salmon must obtain open water cues for orientation. Moreover, lacustrine masu salmon in Lake Toya showed clear diurnal movement when they encountered the mouth of a river for the first time at the beginning of the spawning season. They swam vertically around the thermocline depth in the daytime and stayed at the water surface during the night time. This diurnal movement disappeared gradually toward the peak of the spawning season, and they carried out upstream migration to the spawning ground in rivers. These behavioral changes in masu salmon during the spawning season suggests that masu salmon are able to calculate day length using a biological.

3. Endocrinological Researches on Salmon Imprinting and Homing Migration

Homing migration in salmon is closely related to gonadal maturation, which is regulated mainly by the brain-pituitary-gonadal (BPG) axis. There are two molecular types of gonadotropin-releasing hormone (GnRH), salmon GnRH (sGnRH) and chicken GnRH-II (cGnRH-II), in salmon brains (Amano et al. 1997). In particular, sGnRH in the olfactory system, the terminal nerve, and the preoptic area are considered to play important roles in the homing migration of salmon. Subsequently, sGnRH in the preoptic area controls gonadotropin (GTH), luteinizing hormone (LH) and follicle-stimulating hormone (FSH) synthesis and release from the pituitary gland. GTHs induce steroidogenesis in the gonads, and steroid hormones stimulate gametogenesis and final gameto-maturation; estradiol-17β (E_2) and testosterone (T) are active in vitellogenesis, T and 11-ketotestosterone (11KT) in spermatogenesis, and 17α, 20β-dihydroxy-4-pregnen-3-one (DHP) in final gameto-maturation in both sexes (Nagahama 1999). Hormone profiles have been investigated in the BPG axis of salmon during homing migration and gonadal maturation (Ueda and Yamauchi 1995; Ueda 1999; Ueda 2011; Ueda 2012; Urano et al. 1999; Makino et al. 2007).

3.1. Hormone profiles of chum salmon during homing migration

The hormone profiles in the BPG axis of chum salmon migrating from the Bering Sea to the spawning ground in the Chitose River, Hokkaido, Japan, were measured using specific time-resolved fluoroimmunoassay (TR-FIA) systems developed by Yamada et al. (Yamada et al. 2002). The level of sGnRH in the olfactory bulb (OB) of both sexes showed a peak from the coastal sea to the stream mouth of the Ishikari River where the salmon's olfactory discriminating ability as it enters the the natal stream should be functioning. The level of sGnRH also showed an increase in the telencephalon (TC) at the point where the Chitose River branches from the Ishikari River, an area where the olfactory functions of the chum salmon should also be highly activated. In the pituitary gland, sGnRH levels tended to increase at the same time as elevation in LH levels around the coastal sea in females and the stream mouth of the Ishikari River in males. In contrast, FSH levels did not show any clear correlations with sGnRH levels in the pituitary gland. Although the role of cIIGnRH in these brain regions remains to be elucidated, the levels of cIIGnRH in the medulla oblongata (MO) increased in both sexes at the pre-spawning ground while that in the optic tectum (OT) also increased in males. In the diencephalon (DC) and cerebellum (CB), cIIGnRH levels showed no significant changes during homing migration.

sGnRH immunoreactive neurons, which also showed signals for pro-sGnRH mRNA, were observed in the dorsal portion of the olfactory nerve in chum salmon in the coastal sea, but not in fish at the spawning ground in the natal river (Kudo et al. 1996a). Changes in the levels of GTH subunit mRNAs in the pituitary of pre-spawning chum salmon reported that the levels of GTH α2 and LHβ increased from seawater to freshwater, but those of FSHβ showed no significant changes (Kitahashi et al. 1998b). Serum steroid hormone levels showed similar profiles as previous observations (Ueda et al. 1984; Ueda 1999); E_2 in females and 11KT in males increased during vitellogenesis and spermatogenesis, respectively, and DHP increased dramatically at the time of final gonadal maturation in both sexes. It is interesting to note that both sGnRH levels in the TC and serum T levels in both sexes showed a coincident peak at the branch point of the Chitose River from the Ishikari River. These results confirm that sGnRH plays a role in GTH secretion in the pituitary of chum salmon, and sGnRH and cIIGnRH might be involved in brain region-dependent roles on gonadal maturation and homing migration in salmon. In addition, year-to-year differences in plasma levels of steroid hormones in pre-spawning chum salmon were also observed in relation to sea surface temperature (SST) of the coastal sea, some of which may be influenced by year-to-year variation of SST (Onuma et al. 2003).

3.2. Homing profiles of lacustrine sockeye salmon

In Lake Shikotsu (surface area 78 km^2, average and maximum depth 265 m and 363 m, respectively), adult sockeye salmon were captured from September to November, adjacent to their natal hatchery prior to spawning. They were sampled for serum steroid hormones, tagged, and released in the center of the lake. Fish were sampled again at recapture to characterize changes in steroid hormone levels in individual migrants as well as homing duration and percentage in each month (Sato et al. 1997). Homing duration was significantly shortened from September to October in males and from October to November in females (Fig. 4A). All males returned faster than females early in September and October, although half of the males did not return to the natal site in November. In contrast, 78%–90% of females returned over the entire three month sampling period (Fig. 4B). It is interesting to note that the average homing percentage of both sexes for three months is 83%, indicating no differences in the total number of homing individuals between male and female. Although only old references were available about studying the sex ratio of chum salmon on the spawning ground, Bakkala reported that males predominated early and females late in the spawning run (Bakkala 1970). Although male salmonids do not show any territorial behaviors, they maintain high levels of aggressive behavior to

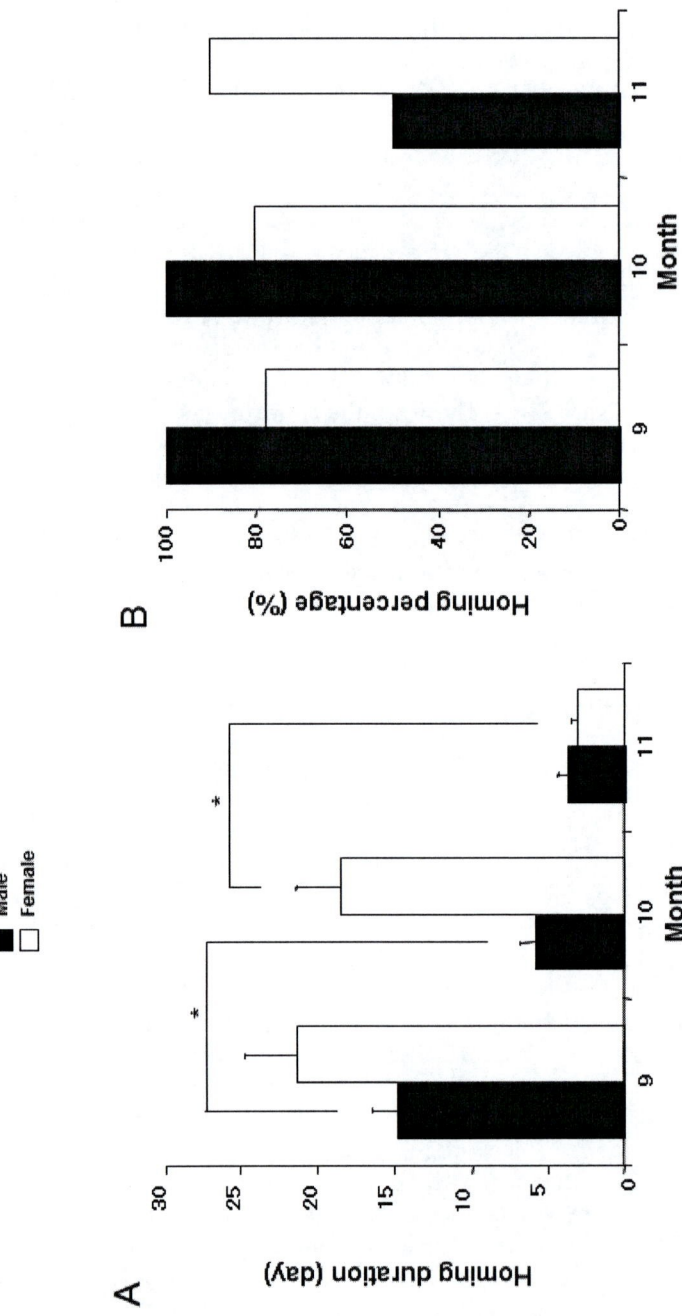

Fig. 4. Changes in homing duration (A) and percentage (B) of lacustrine sockeye salmon in Lake Shikotsu from September to November. Significant differences at 5% (*) level is indicated.

compete for access to females (Jones 1959), suggesting that early returning males might accrue some benefits in securing females for breeding.

The differences between the sexes in homing behavior are thought to be reflected by the different steroid hormone profiles between males and females (Sato et al. 1997). In males, the shortening of homing duration coincided with an increase in serum T and 11KT levels. The reduction of homing percentage was associated with decreased serum T levels and increased serum DHP levels. In females, the shortening of homing duration corresponded to an elevation of serum T and DHP levels, and a drop in serum E_2 levels.

3.3. Hormonal manipulation in sockeye salmon

Since GnRH treatment has been reported to be highly effective in inducing GTH release, ovulation and spermiation in teleost fishes (Zohar 1996), we investigated the effect of GnRH analog (GnRHa) implantation on both homing profiles and serum steroid hormone levels of fish in September (Sato et al. 1997; Kitahashi et al. 1998c). The GnRHa implantation was highly efficient in shortening the homing duration, and caused dramatic increases in serum DHP levels in both sexes. An interesting discrepancy was observed between rapidly and slowly returning individual males: rapidly returning males showed higher serum T levels and lower serum DHP levels than slowly returning individual males. To examine the direct action of T and DHP on homing duration, T and DHP were implanted in fish in September in comparison with GnRHa-implantation (Kitahashi et al. 1998c). GnRHa-implanted fish returned significantly earlier than the control fish regardless of sex. T implantation tended to reduce homing duration in both males and females, but there was no statistical significance. DHP implantation also significantly shortened homing duration in females, but it did not have any significant effect in males (Fig. 5A). These steroid hormone implantations did not affect serum T and DHP levels. It is interesting to note that the direct actions of T and DHP on homing migration are sex dependent. The peak of plasma T levels in lacustrine sockeye salmon of both sexes was observed at the time when they gathered at the mouth of their natal stream in Lake Chuzenji, Japan (Ikuta 1996). Androgens are well-known to be involved in stimulating aggressive behavior in teleost fishes (Villars 1983), and serum T and 11KT are the two major androgens which influence spawning behaviors, downstream and upstream migration, and the social dominance hierarchy (Kindler et al. 1989; Cardwell and Liley 1991; Pankhurst and Barnett 1993; Brantly et al. 1993; Cardwell et al. 1996; Munakata et al. 2001a; Munakata 2001b). Although DHP is known to be a maturation-inducing steroid in salmonids (Nagahama and Adachi 1985), its function to the central nervous system has not yet been clarified. The functional roles of T and DHP on the

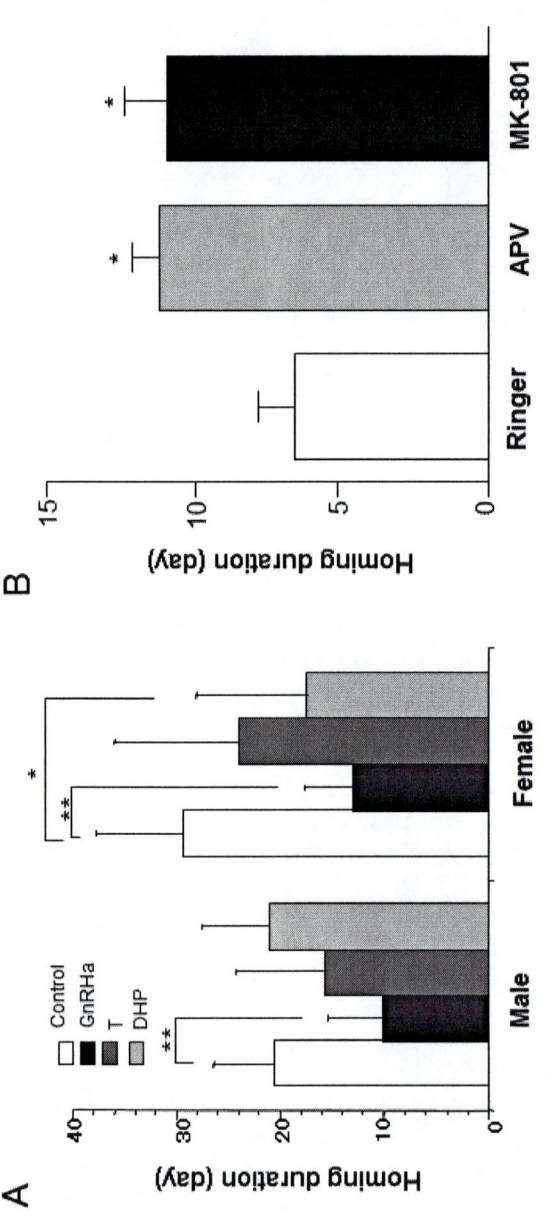

Fig. 5. (A) Effects of GnRH analog (GnRHa), testosterone(T) and 17α,20β-dihydroxy-4-pregnen-3-one (DHP) implantation on homing duration of lacustrine sockeye salmon in Lake Shikotsu in September. (B) Effects of MNDA blockers (APV and MK-801) on homing duration of male lacustrine sockeye salmon in Lake Shikotsu in October. Significant differences at 5% (*) and 1% (**) levels are indicated.

salmonid homing migration should be further investigated with special attention to their action on the central nervous system.

GnRHa implantation was also highly effective in accelerating gonadal maturation in anadromous, maturing sockeye salmon of both sexes. Expression of GTH subunit genes in the pituitary gland was examined and revealed that the levels of GTH α and LHβ mRNAs in GnRHa-implanted fish were higher than those in control fish, but the levels of FSHβ mRNA showed no change (Kitahashi et al. 1998a). Implantation of GnRHa caused a significant elevation of serum DHP levels in both sexes, but had no effect on levels of T and 11KT in males or E_2 and T in females (Fukaya et al. 1998). These data suggest that sGnRH in the brain stimulates LH release from the pituitary gland, and then LH enhances serum DHP levels in both sexes during the latter part of the homing migration in salmonid fishes. GnRH is believed to play a leading role in the homing migration of both sexes, but gonadal steroids, especially T and DHP, seem to have sexually different influences on homing migration.

Recently, we have applied blood oxygenation level-dependent (BOLD) functional magnetic resonance imaging (fMRI) to investigate the odor information processing of the natal stream in the brain of lacustrine sockeye salmon, and found that strong responses to odors of the natal stream were mainly observed in the lateral area of dorsal telencephalon (Dl), which are homologous to the medial pallium (hippocampus) in terrestrial vertebrates (Bandoh et al. 2011). Olfactory memory plays a key role in imprinting and recalling natal stream odor information in salmon. In the formation of memory, the possible role of long-term potentiation (LTP) has been studied with a focus on N-methyl-D-aspartate (NMDA) receptors, which induce LTP (Martin et al. 2000). LTP is known to occur in the brain of zebrafish (*Danio rerio*) (Nam et al. 2004), rainbow trout (*O. mykiss*) (Kinoshita et al. 2005), and common carp (*Cyprinus carpio*) (Satou et al. 2006). Effects of NMDA receptor blockers (APV and MK-801) on homing duration of male lacustrine sockeye salmon in Lake Shikotsu in late October were investigated, and revealed that homing duration was significantly prolonged by both blockers (Fig. 5B). These results suggest that NMDA receptors might be deeply involved in recalling the imprinting memory in sockeye salmon, and the exact roles of NMDA during imprinting migration should be investigated.

3.4. Hormone profiles of salmon during imprinting migration

The imprinting migration in salmon is considered to be closely implicated in smoltification, which is regulated mainly by the brain-pituitary-thyroidal (BPT) axis, but hormone profiles in the BPT axis have not been examined deeply. In mammals, thyrotropin-releasing hormone (TRH) in the hypothalamus controls thyroid-stimulating hormone (TSH) synthesis and release from the pituitary gland, and then TSH stimulates the release of thyroxine (T_4) and triiodothyronine (T_3) from the thyroid gland. TRH may also function as a neurotransmitter and/or neuromodulator in the central nervous system and peripheral organs (Jackson and Reichlin 1974). In the brain of sockeye salmon, mRNA encoding TRH precursor (Ando et al. 1998) and cDNA encoding TRH receptor subtypes (Saito et al. 2011) have been reported. Also, T_4 surge was observed during smoltification in salmonid fishes (Dickhoff et al. 1982; Yamauchi et al. 1984; Dickhoff and Sullivan 1987). Hormone profiles TRH, TSH, and thyroid hormones in the BPT axis of salmon should be investigated during imprinting migration in detail.

4. Neurophysiological Researches on Salmon Olfactory Imprinting and Homing Abilities

Since the olfactory hypotheses for salmon imprinting and homing to their natal stream was proposed by Hasler's research group in 1950s (Hasler and Wisby 1951; Wisby and Hasler 1954), mechanisms of olfactory imprinting and homing abilities in salmon have been intensively studied (Hasler and Scholz 1983; Døving 1989; Stabell 1992; Dittman and Quinn 1996; Bertmar 1997; Nevitt and Dittman 1998; Quinn 2005; Ueda et al. 2007; Hino et al. 2009; Ueda 2011; Ueda 2012). The pheromone hypothesis proposed by Nordeng (Nordeng 1971; Nordeng 1977) using Arctic char (*Salvelinus alpines*) and Atlantic salmon (*Salmo salar*) suggested that juvenile salmon in a stream released population-specific odors that guided homing adults. Several studies have also suggested that juvenile salmonids produce population-specific odors or pheromones (Groot et al. 1986; Quinn and Tolson 1986; Courtenay et al. 1997). It has also been demonstrated that sex steroids and prostaglandins that have effects on the olfactory epithelium of salmonids may be acting as sexual pheromones (Moore and Scott 1992; Moore and Warning 1996). Recently, L-kynurenine, an amino acid was identified as a sex pheromone in the urine of ovulated female masu salmon (Yambe et al. 2006). However, there are no juveniles of chum salmon or pink salmon present at the time that the adults return. Nonetheless, it is now widely accepted that some specific odors in the natal stream are important for olfactory imprinting and homing in salmon.

4.1. Electrophysiological studies on olfactory discriminating ability

We examined the olfactory discriminatory ability of lacustrine sockeye and masu salmon, which were reared in the culture pond at Toya Lake Station, by recording the integrated olfactory nerve response (Sato et al.

2000). The olfactory organs of both species elicited different responses to various freshwaters, regardless of sex or gonadal maturity. The source and effluent water from the culture pond evoked the minimum and maximum responses respectively. These odors may modify the source water in such a way as to make the culture pond water more detectable to the olfactory system. In cross-adaptation experiments, the stream waters abolished the secondary response to the lake water, but the lake water did not abolish the secondary response to the stream waters. This phenomenon is quite reasonable because the salmon migrate from the lake to the stream. The minimum concentration (threshold) required to induce the olfactory nerve response to the culture pond water after adaptation to the lake water was between 0.1% and 1.0%. This threshold level suggests that the olfactory discriminatory ability of salmonids during homing migration must function within a limited distance from the natal stream.

4.2. Properties of natal stream odors

Several attempts to identity the natal stream odor were made based on the olfactory bulbar response, and suggested that the natal stream odors were non-volatile (Fagerlund et al. 1963; Cooper et al. 1974; Bodznick 1978). Spectral analysis of the olfactory bulbar response suggested that the natal stream odor was absorbed on activated carbon and ion-exchange resin, insoluble in petroleum-ether, dialyzable, non-volatile, and heat-stable (Ueda 1985). Unlike olfactory organs of terrestrial animals, fish olfactory organs respond only to a limited number of chemicals dissolved in water. Chemicals that elicit the response from the olfactory organs of salmon are amino acids, steroids, bile acids, and prostaglandins (Hara 1994).

We analyzed the compositions of dissolved free amino acids (DFAA), inorganic cations and bile acids in waters from three streams which flow into Lake Toya (Shoji et al. 2000). Application of mixtures of inorganic cations or bile acids to the olfactory epithelium, based on their compositions in stream waters, induced only very small responses. On the other hand, application of mixtures of DFAA induced large responses. The response to artificial stream water, based on the composition of DFAA and salts, closely resembled the response to the corresponding natural stream water. Cross-adaptation experiments with three combinations of natural and artificial stream waters were carried out. The response pattern for each combination of artificial stream water closely resembled that of the corresponding combination of natural stream water. Accordingly, we proposed that DFAA compositions in the natal stream water are likely natal stream odors.

Changes in the DFAA compositions in stream water are attributed mainly to complicated biological processes in the watershed ecosystem. There are many possible factors affecting the DFAA compositions both within and beyond the stream environment, such as soils, vegetation, litter, pollen, dew, and various microbial activities (Thomas 1997). Among these factors, the roles of complex microbial communities called biofilms have been intensively investigated (Costerton et al. 1994; Nosyk et al. 2008). A biofilm consists of various microorganisms, and is embedded into a matrix of extracellular polymeric substances. We investigated the origin of DFAA in stream water, focusing on biofilms in the stream bed by means of incubation experiments in the laboratory. Stones were placed in the Toyohira Stream, Hokkaido, for three months, allowing formation of biofilms, and then incubated for 24 hours in the laboratory at stream water temperature. After incubation, the composition and concentrations of DFAA in the incubation solution were measured by a high-performance liquid chromatography (HPLC). The DFAA concentration increased greatly in the biofilm incubation solution of the treatment group, but the DFAA composition (mole %) showed little change after 24 hr incubation, which was similar to stream water. These results suggest that biofilms are a major source of DFAA in stream water (Ishizawa et al. 2010).

4.3. Behavioral studies on olfactory discriminating abilities

Behavior experiments were carried out to test attractant effects on upstream selective movement among four Pacific salmon (pink, chum, sockeye, and masu salmon), using artificial natal stream water (ANW) prepared to the same composition and concentration of DFAA in their natural natal stream in a two-choice test tank (Y-maze) consisting of two water inlet arms and one pool. Either ANW or natural lake water (NLW) was added to the water inlet of either left or right arm and the fish movement monitored to determine the number of fish that moved to each arm. In combination of ANW and NLW, pink salmon showed the highest percentage of upstream movement among the four Pacific salmon species, but showed the lowest percentage of selectivity in the arm running ANW (Ueda 2011). These results indicated that ANW had different attractant effects on selective upstream movement among the four Pacific salmon species. Pink salmon showed the highest upstream movement but showed the lowest selectively to the artificial natal stream water. It is interesting to note the evolutionary relationship between the olfactory discriminating ability and the homing accuracy among four Pacific salmon. If salmon conduct very accurate homing migrations to their natal stream, there would be little chance to expand their distribution area, which would in turn affect population size and genetic diversity. Thus pink salmon may have evolved the capacity to select non-natal stream odors during homing migration.

Further behavioral experiments with chum salmon captured in the Osaru River (OR), Hokkaido, were also conducted in the Y-maze using various combinations of control water (NLW) and three artificial stream waters prepared by using the same composition and concentration of DFAA found in natural stream waters: 1. artificial OR water (AOR) 2. AOR without L-glutamic acid, the major amino acid in OR water (AOR-E) 3. Another artificial water (ALS) had much higher amino acid concentrations than OR (Yamamoto and Ueda 2009). In behavioral tests, the fish did not discriminate between AOR and AOR-E, but displayed significant selection of AOR or AOR-E over NLW and AOR over ALS. Electrophysiological cross-adaptation experiments indicated that mature male chum salmon have the olfactory capability to distinguish between AOR and AOR-E. These results suggest that migratory male chum salmon respond to DFAA mixtures in their natal stream water and appear not to be affected by single amino acids.

4.4. Artificial imprinting studies using amino acids

By using artificial odors, β-phenylethyl alcohol (PEA) or morpholine, coho salmon that had been imprinted with these odors during smoltification were lured into unfamiliar streams scented with these odors during homing migration a few years later (Cooper et al. 1976; Scholz et al. 1976). The olfactory receptor cells of coho salmon that had been imprinted with PEA had a higher sensitivity to PEA as compared with non-imprinted fish (Nevitt et al. 1994), and only fish that were exposed to PEA or natural stream odors during smoltification formed an imprinted memory (Dittman et al. 1996). Using electrophysiological and behavioral experiments, we have revealed that one-year-old lacustrine sockeye salmon can be imprinted around the stage of smoltification by a single amino acid, 1 μM L-proline (Pro) or L-glutamic acid (Glu). The electro-olfactogram (EOG) responses of test fish exposed to Pro in March (before smoltification) and April–June (during smoltification) for two weeks were significantly greater than those of non-exposed control fish, but not those of test fish exposed in July (after smoltification). When Pro and control water were added to the water inlets of the Y-maze during the spawning season two years after the test water exposure, 80% of maturing and matured test fish exposed before and during smoltification showed a preference for Pro, whereas those exposed after smoltification did not (Fig. 6). The EOG response of test fish exposed to Pro or Glu for one hour, six hours, one day, seven days, or 14 days in May revealed that only the response after 14-day exposure was significantly greater than the control. We conclude that one-year-old lacustrine sockeye salmon can be imprinted by a single amino acid before and during smoltification, and that imprinting requires exposure for at least 14 days (Yamamoto et al. 2010).

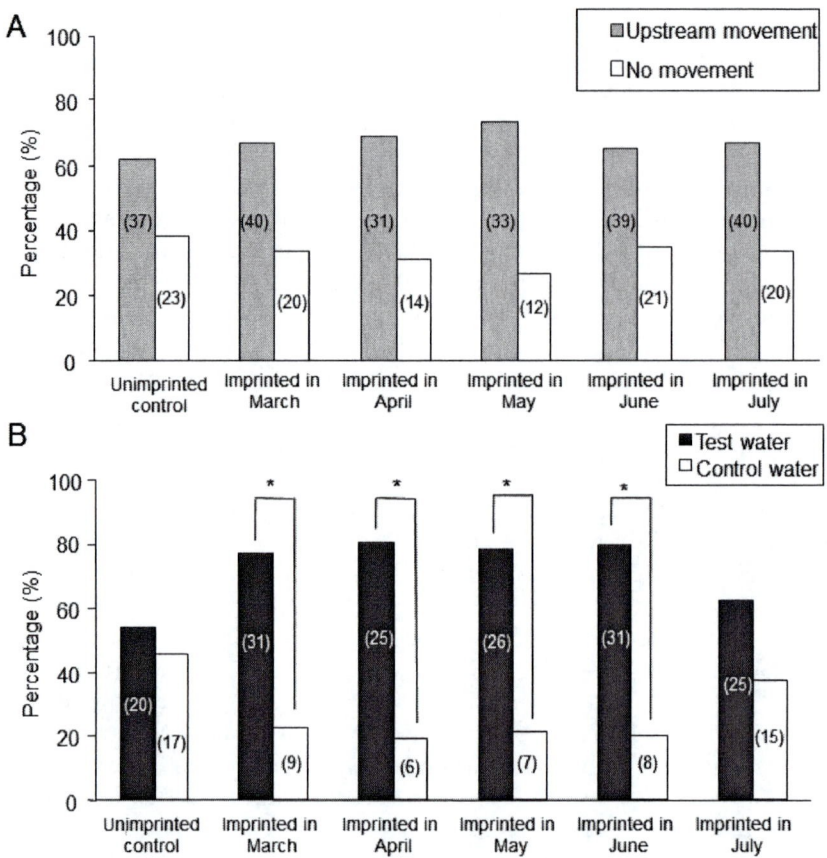

Fig. 6. Upstream movement (A) and selectivity (B) of mature male lacustrine sockeye salmon (experimental fish exposed to 1 µM L-proline (Pro) from March to July 2 years previously and unimprinted control fish) in the two-choice test tank containing either test water (Pro) or control water (Toya Lake Water). Numbers in parenthesis indicate the number of fish. Significant differences at 5% (*) levels are indicated.

4.5. Biochemical and molecular biological studies on salmon olfactory functions

Using sodium dodecyl sulfate-polyacrylamide gel electrophoresis, an olfactory system-specific protein of 24 kDa (N24) was identified in lacustrine sockeye salmon by electrophoretic comparison of proteins restricted to the olfactory system with those found in other parts of the brain (Shimizu et al. 1993). In various species of teleosts, N24 immunoreactivity was found in the olfactory system of species migrating between the sea and freshwater streams, such as Japanese eel (*Anguilla japonica*), but not in non-migratory species, such as carp (*Cyprinus carpio*) (Ueda et al. 1994). Interestingly, N24 immunoreactivity was also observed in the testicular germ cells, spermatids and spermatozoa, suggesting its involvement in sperm chemotaxis (Ueda et al. 1993). Immunocytochemical and immunoelectronmicroscopic observations revealed that N24 positive immunoreactivity occurred in ciliated and microvillous olfactory receptor cells and the glomerular layer near the mitral cells in the olfactory bulb (Kudo et al. 1996b; Yanagi et al. 2004). cDNA encoding N24 was isolated and sequenced, and this cDNA contained a coding region encoding 216 amino acid residues. Protein and nucleotide sequencing demonstrated the existence of a remarkable homology between N24 and glutathione S-transferase class pi enzymes (Kudo et al. 1999). Northern analysis showed that N24 mRNA with a length of 950 bases was expressed in lacustrine sockeye salmon olfactory epithelium. The functional roles of N24 during salmon homing migration are still unclear, but N24 is a useful molecular marker for studying olfactory functions in salmonids.

Salmon olfactory imprinting-related gene (SOIG) from the olfactory system of lacustrine sockeye salmon has been identified by subtractive hybridization technique of cDNA-representational difference analysis (cDNA-RDA) using fish at smoltification as a tester and fish at the feeding migration term as a driver (Hino et al. 2007). SOIG mRNA was shown to be expressed in olfactory receptor cells and basal cells of the olfactory epithelium. The expression levels of SOIG mRNA in the olfactory epithelium have been analyzed during several lifecycle stages of lacustrine sockeye salmon and chum salmon, such as ontogeny, smoltification, and homing. During ontogeny, the expression levels of SOIG mRNA in chum salmon were significantly higher in alevin (juvenile fry) than in embryos at 43 and 60 days after fertilization. During smoltification, SOIG mRNA levels increased before and during smoltification, and decreased after smoltification in sockeye salmon (Fig. 7A). These changes were coincided with serum thyroxine changes during smoltification (Fig. 7B) (Yamamoto et al. 2010). During homing migration, SOIG mRNA levels in the olfactory epithelium of chum salmon were elevated at the estuary and pre-spawning ground. It is thought that SOIG might be related to olfaction or cell proliferation during both smoltification and the final stage of homing.

The olfactory chemoreception is accomplished through binding of the odorant substance to an olfactory receptor (OR) that is reportedly encoded by 100–200 genes (Alioto and Ngai 2005) in the olfactory epithelium with subsequent propagation of the information to the central nervous system. There are two types of OR genes namely, main olfactory receptors (MORs), which are expressed in ciliated olfactory receptor cells; and vomeronasal olfactory receptors (VORs, subdivided into V1R and V2R), which are expressed in microvillous olfactory receptor cells. MOR genes have also been identified in a number of salmonids (Wickens et al. 2001; Dukes et al. 2004, 2006; Morinishi et al. 2007). Recently, olfactory receptor expression

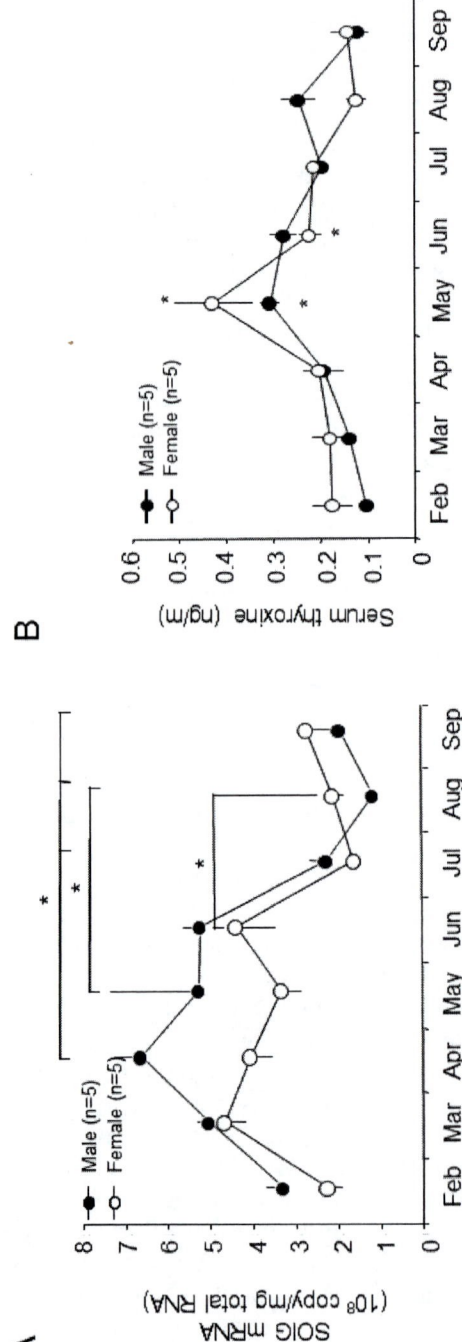

Fig. 7. (A) Changes in salmon olfactory imprinting-related gene (SOIG) mRNA expression levels of one-year-old lacustrine sockeye salmon from February to September. (B) Changes in serum thyroxine (T_4) levels of one-year-old lacustrine sockeye salmon from February to September. Significant differences at 5% (*) levels are indicated.

was investigated in different life stages of Atlantic salmon, demonstrating that seven V2R-like (*OlfC*) genes were expressed at higher levels in juveniles (parr and smolts) than in adults (Johnstone et al. 2011). Although many MORs and VORs have been identified from several vertebrates owing to the progress of whole genome analysis, many ligands remain uncharacterized. Further intensive molecular biological researches are needed to clarify the olfactory chemoreception during imprinting and homing migration in salmon.

Conclusion

This review describes recent studies on physiology of imprinting and homing migration mainly in anadromous chum salmon from the Bering Sea to Japan as well as lacustrine sockeye salmon in Lake Toya and Lake Shikotsu. Using these model fish, three different approaches (physiological biotelemetry researches on salmon homing migration, endocrinological researches on salmon imprinting and homing migration, and neurophysiological researches on salmon olfactory imprinting and homing ability) have provided valuable new understanding of the physiology of imprinting and homing migration in Pacific salmon. However, many questions remain unanswered, such as the sensory mechanisms of open water orientation, the hormonal control mechanisms for sensory systems and the central nervous system, accurate and false homing, and seasonal and yearly changes in DFAA composition in natal stream water. Despite the difficulties of a temporally limited spawning season, research from molecular biology to behavioral biology will provide new concepts for physiology of imprinting and homing migration in salmon.

Acknowledgements

I would like to thank the following collaborative researchers and students in my laboratory for their valuable contributions to the present study:

Behavioral research:

M. Kaeriyama, Y. Naito, H. Tanaka, H. Sakano, J.B.K. Leonard, A. Sato, K. Orito, E. Fujiwara, Y. Matsushita, M. Akita, H. Nii, Y. Makiguchi, K. Hayashida, and K. Miyoshi.

Endocrinal research:

K. Yamauchi, M. Amano, H. Yamada, M. Iwata, A. Urano, S. Hyodo, Y. Zohar, D. Alok, M. Ban, S. Taniyama, S. Matsumoto, R.K. Bhandari, T. Kitani, S.P. Lee, K. Fukaya, N. Furukawa, and R. Murakami.

Olfactory analyses:

T. Shoji, K. Kurihara, H. Nagasawa, H. Kudo, M. Shimizu, S. Yanagi, K. Shimozawa, K. Sato, M. Fukaya, H. Hino, N. Ileva, Y. Yamamoto, and S. Ishizawa.

The present study was supported in part by Grant-in-Aid for Scientific Research from the Ministry of Education, Culture, Sports, Science and Technology (MEXT), Japan, from the Japan Society for the Promotion of Science (JSPS), from the Hokkaido Foundation for the Promotion of Scientific and Industrial Technology, from the Mitsubishi Foundation, from Mitsui & Co. Ltd., and from the Hokkaido University.

References

Alioto, T.S. and J. Ngai. 2005. The odorant receptor repertoire of teleost fish. BMC Genomics 6: 173.

Amano, M., A. Urano, and K. Aida. 1997. Distribution and function of gonadotropin-releasing hormone (GnRH) in the teleost brain. Zool. Sci. 14: 1–11.

Ando, A., J. Ando, and A. Urano. 1998. Localization of mRNA encoding thyrotropin-releasing hormone precursor in the brain of sockeye salmon. Zool. Sci. 15: 945–953.

Bakkala, R.G. 1970. Synopsis of biological data on the chum salmon, *Oncorhynchus keta* (Walbaum) 1792. FAO Fish Synopsis 41; US Fish and Wildlife Service Circular 315: 1–89.

Bandoh, H., I. Kida, and H. Ueda. 2011. Olfactory responses to natal stream water in sockeye salmon by BOLD fMRI. PLoS ONE 6: e16051.

Bertmar, G. 1997. Chemosensory orientation in salmonid fishes. Advan. Fish Res. 2: 63–80.

Bodznick, D. 1978. Calcium ion: An odorant for natural water discriminations and the migratory behavior of sockeye salmon. J. Comp. Physiol. 127: 157–166.

Brantly, R.K., J.C. Wingfield, and A.H. Bass. 1993. Sex steroid levels in *Porichthys notatus*, a fish with alternative reproductive tactics, and a review of the hormonal bases for male dimorphism among teleost fishes. Horm. Behav. 27: 332–347.

Cardwell, J.R. and N.R. Liley. 1991. Androgen control of social status in males of a wild population of stoplight parrotfish, *Sparisome viride* (Scaridae). Horm. Behav. 25: 1–18.

Cardwell, J.R., P.W. Sorensen, G.J. Van Der Kraak, and N.R. Liley. 1996. Effect of dominance status on sex hormone levels in laboratory and wild-spawning male trout. Gen. Comp. Endocrinol. 101: 333–341.

Cooke, S.J., S.G. Hinch, M. Wikelski, R.D. Andrews, L.J. Kuchel, T.G. Wolcott, and P.J. Butler. 2004. Biotelemetry: a mechanistic approach to ecology. Trends Ecol. Evol. 19: 334–343.

Cooper, J.C., G.F. Lee, and A.E. Dizon. 1974. An evaluation of the use of the EEG technique to determine chemical constituents in homestream water. Wis. Acad. Sci. Arts Lett. 62: 165–172.

Cooper, J.C., A.T. Scholz, R.M. Horrall, A.D. Hasler, and D.M. Madison. 1976. Experimental confirmation of the olfactory hypothesis with artificially imprinted homing coho salmon (*Oncorhynchus kisutch*). J. Fish. Res. Board Can. 33: 703–710.

Costerton, J.W., Z. Lewandowski, D. DeBeer, D. Caldwell, D. Korber, and G. James. 1994. Biofilms, the customized microniche. J. Bacteriol. 176: 2137–2142.

Courtenay, S.C., T.P. Quinn, H.M.C. Dupuis, C. Groot, and P.A. Larkin. 1997. Factors affecting the recognition of population-specific odors by juvenile coho salmon. J. Fish Biol. 50: 1042–1060.

Dickhoff, W.W., D.S. Darling, and A. Gorbman. 1982. Thyroid function during smoltification of salmonid fish. Gunma Symp. Endocrinol. 19: 45–61.

Dickhoff, W.W. and C. Sullivan. 1987. Involvement of the thyroid gland in smoltification, with special reference to metabolic and developmental processes. Am. Fish. Soc. Symp. 1: 197–210.

Dittman, A.W. and T.P. Quinn. 1996. Homing in Pacific salmon: Mechanisms and ecological basis. J. Exp. Biol. 199: 83–91.

Dittman, A.W., T.P. Quinn, and G.A. Nevitt. 1996. Timing of imprinting to natural and artificial odors by coho salmon (*Oncorhynchus kisutch*). Can. J. Fish. Aqua. Sci. 53: 434–442.

Døving, K.B. 1989. Molecular cues in salmonid migration. *In:* Maruani, J. [ed.]. Molecules in Physics, Chemistry, and Biology. Kluwer Academic Publishers, Amsterdam, Netherlands pp. 299–329.

Døving, K.B., H. Westerberg, and P.B. Johnsen. 1985. Role of olfaction in the behavioral and neuronal responses of Atlantic salmon, *Salmo salar*, to hydrographic stratification. Can. J. Fish. Aquat. Sci. 42: 1658–1667.

Dukes, J.P., R. Deaville, M.W. Bruford, A.F. Youngson, and W.C. Jordan. 2004. Odorant receptor gene expression changes during the parr-smolt transformation in Atlantic salmon. Mol. Ecol. 13: 2851–2857.

Dukes, J.P., R. Deaville, D. Gottelli, J.E. Neigel, M.W. Bruford, and W.C. Jordan. 2006. Isolation and characterisation of main olfactory and vomeronasal receptor gene families from the Atlantic salmon (*Salmo salar*). Gene 371: 257–267.

Fagerlund, U.H.M., J.R. McBridge, M. Smith, and N. Tomlinson. 1963. Olfactory perception in migrating salmon III. Stimulants for adult sockeye salmon (*Oncorhynchus nerka*) in home stream waters. J. Fish. Res. Board Can. 20: 1457–1463.

Fukaya, M., H. Ueda, A. Sato, M. Kaeriyama, H. Ando, Y. Zohar, A. Urano, and K. Yamauchi. 1998. Acceleration of gonadal maturation in anadromous maturing sockeye salmon by gonadotropin-releasing hormone analog implantation. Fish. Sci. 64: 948–951.

Groot, C. and L. Margolis 1991. Pacific Salmon Life Histories. UBC Press, Vancouver. Canada.

Groot, C., T.P. Quinn, and T.J. Hara. 1986. Responses of migrating adult sockeye salmon (*Oncorhynchus nerka*) to population-specific odors. Can. J. Zool. 64: 926–932.

Hansen, L.P., N. Jonsson, and B. Jonsson. 1993. Oceanic migration in homing Atlantic salmon. Anim. Behav. 45: 927–941.

Hara, T.J. 1994. The diversity of chemical stimulation in fish olfaction and gustation. Rev. Fish Biol. Fish. 4: 1–35.

Hasler, A.D. and A.T. Scholz. 1983. Olfactory imprinting and homing in salmon. Springer-Verlag, New York. USA.

Hasler, A.D. and W.J. Wisby. 1951. Discrimination of stream odors by fishes and relation to parent stream behavior. Am. Natural. 85: 223–238.

Hino, H., T. Iwai, M. Yamashita, and H. Ueda. 2007. Identification of an olfactory imprinting-related gene in the lacustrine sockeye salmon, *Oncorhynchus nerka*. Aquacult. 273: 200–208.

Hino, H., N.G. Miles, H. Bandoh, and H. Ueda. 2009. Molecular biological research on olfactory chemoreception in fishes. J. Fish Biol. 75: 945–959.

Ikuta, K. 1996. Effects of steroid hormones on migration of salmonid fishes. Bull. Nat. Res. Inst. Aquacult. Suppl. 2: 23–27.

Ishizawa, S., Y. Yamamoto, T. Denboh, and H. Ueda. 2010. Release of dissolved free amino acids from biofilms in stream water. Fish. Sci. 76: 669–676.

Jackson, I.M. and S. Reichlin. 1974. Thyrotropin-releasing hormone (TRH): distribution in hypothalamic and extrahypothalamic brain tissues of mammalian and submammalian chordates. Endocrinol. 95: 8548–8562.

Johnstone, K.A., K.P. Lubienieck, B.F. Koop, and W.S. Davidson. 2011. Expression of olfactory receptors in different life stages and life histories of wild Atlantic salmon (*Salmon salar*). Mol. Ecol. 20: 4059–4069.

Jones, J.W. 1959. The Salmon. Collines Clear-Type Press, London. UK.

Kaeriyama, M. and H. Ueda. 1998. Life history strategy and migration pattern of juvenile sockeye (*Oncorhynchus nerka*) and chum salmon (*O. keta*) in Japan: a review. North Pac. Anadr. Fish Comm. Bull. 1: 163–171.

Kindler, P.W., D.P. Philipp, M.T. Gross, and J.M. Bahr. 1989. Serum 11-ketotestosterone and testosterone concentrations associated with reproduction in male bluegill (*Lepomis macrochirus*: Centrarchidae). Gen. Comp. Endocrinol. 75: 446–453.

Kinoshita, M., M. Fukaya, T. Tojima, S. Kojima, H. Ando, and M. Watanabe. 2005. Retinotectal transmission in the optic tectum of rainbow trout. J. Comp. Neurol. 484: 249–259.

Kitahashi, T., D. Alok, H. Ando, M. Kaeriyama, Y. Zohar, H. Ueda, and A. Urano. 1998a. GnRH analog stimulate gonadotropin II gene expression in maturing sockeye salmon. Zool. Sci. 15: 761–765.

Kitahashi, T., H. Ando, M. Ban, H. Ueda, and A. Urano. 1998b. Changes in the levels of gonadotropin subunit mRNAs in the pituitary of pre-spawning chum salmon. Zool. Sci. 15: 753–760.

Kitahashi, T., A. Sato, D. Alok, M. Kaeriyama, Y. Zohar, K. Yamauchi, A. Urano, and H. Ueda. 1998c. Gonadotropin-releasing hormone analog and sex steroids shorten homing duration of sockeye salmon in Lake Shikotsu. Zool. Sci. 15: 767–771.

Kudo, H., S. Hyodo, H. Ueda, O. Hiroi, K. Aida, A. Urano, and K. Yamauchi. 1996a. Cytophysiology of gonadotropin-releasing-hormone neurons in chum salmon (*Oncorhynchus keta*) forebrain before and after upstream migration. Cell Tiss. Res. 284: 261–267.

Kudo, H., H. Ueda, and K. Yamauchi. 1996b. Immunocytochemical investigation of salmonid olfactory system-specific protein in the kokanee salmon (*Oncorhynchus nerka*). Zool. Sci. 13: 647–653.

Kudo, H., H. Ueda, K. Mochida, S. Adachi, A. Hara, H. Nagasawa, Y. Doi, S. Fujimoto, and K. Yamauchi. 1999. Salmonid olfactory system-specific protein (N24) exhibits glutathione S-transferase class pi-like structure. J. Neurochem. 72: 1344–1352.

Makino, K., T. Onuma, T. Kitahashi, H. Ando, M. Ban, and A. Urano. 2007. Expression of hormone genes and osmoregulation in homing chum salmon: A minireview. Gen. Comp. Endocrinol. 152: 304–309.

Martin, S.J., P.D. Grimwood, and R.G. Morris. 2000. Synaptic plasticity and memory: an evaluation of the hypothesis. Ann. Rev. Neurosci. 23: 649–711.

Moore, A. and A.P. Scott. 1992. 17α,20β-dihydroxy-4-pregnen-3-one 20-sulphate is a potent odorant precocious male Atlantic salmon (*Salmo salar* L.) parr which have been pre-exposed to the urine of ovulated females. Proc. Roy. Soc. London B. 249: 205–209.

Moore, A. and C.P. Warning. 1996. Electrophysiological and endocrinological evidence that F-series prostaglandins function as priming pheromones in mature male Atlantic salmon (*Salmo salar*) parr. J. Exp. Biol. 199: 2307–2316.

Morinishi, F., T. Shiga, N. Suzuki, and H. Ueda. 2007. Cloning and characterization of an odorant receptor in five Pacific salmon. Comp. Biochem. Physiol. 148B: 329–336.

Munakata, A., M. Amano, K. Ikuta, S. Kitamura, and K. Aida. 2001a. The involvement of sex steroid hormones in downstream and upstream migratory behavior of masu salmon. Comp. Biochem. Physiol. 129B: 661–669.

Munakata, A., M. Amano, K. Ikuta, S. Kitamura, and K. Aida. 2001b. The effects of testosterone on upstream migratory behavior in masu salmon, *Oncorhynchus masou*. Gen. Comp. Endocrinol. 122: 329–340.

Murata, S., N. Takasaki, M. Saitoh, H. Tachida, and N. Okada. 1996. Details of retropositional genome dynamics that provide a rationale for a generic division: the distinct branching of all the Pacific salmon and trout (*Oncorhynchus*) from the Atlantic salmon and trout (*Salmo*). Genet. 142: 915–926.

Nagahama, Y. 1999. Gonadal steroid hormones: Major regulators of gonadal differentiation and gametogenesis in fish. *In*: B. Norberg, O.S. Kjesbu, G.L. Taranger, E. Andersson, and S.O. Stefansson [eds.]. Proceedings of the 6th International Symposium on the Reproductive Physiology of Fish, Bergen, Norway. pp. 211–222.

Nagahama, Y. and S. Adachi. 1985. Identification of a maturation-inducing steroid in a teleost, the amago salmon (*Oncorhynchus rhodurus*). Develop. Biol. 109: 428–435.

Nam, R.H., W. Kim, and C.J. Lee. 2004. NMDA receptor-dependent long-term potentiation in the telencephalon of the zebrafish. Neurosci. Lett. 370: 248–251.

Nevitt, G.A. and A.H. Dittman. 1998. A new model for olfactory imprinting in salmon. Integrat. Biol. 1: 215–223.

Nevitt, G.A., A.H. Dittman, T.P. Quinn, and W.J.Jr. Moody. 1994. Evidence for a peripheral olfactory memory in imprinted salmon. Proc. Nat. Acad. Sci. USA 91: 4288–4292.

Nordeng, H.A. 1971. Is the local orientation of anadromous fishes determined by pheromones? Nature. 233: 411–413.

Nordeng, H.A. 1977. A pheromone hypothesis for homeward migration in anadromous salmonids. Oikos. 28: 155–159.

Nosyk, O., E.T. Haseborg, U. Metzger, and F.H. Frimmel. 2008. A standardized pre-treatment method of biofilm flocs for fluorescence microscopic characterization. J. Microbiol. Meth. 75: 449–456.

Ogura, M. and Y. Ishida. 1994. Homing behavior and vertical movements of four species of Pacific salmon (*Oncorhynchus* spp.) in the central Bering Sea. Can. J. Fish. Aquat. Sci. 52: 532–540.

Onuma, T., Y. Higashi, H. Ando, M. Ban, H. Ueda, and A. Urano. 2003. Year-to-year differences in plasma levels of steroid hormones in pre-spawning chum salmon. Gen. Comp. Endocrinol. 133: 199–215.

Pankhurst, N.W. and C.W. Barnett. 1993. Relationship of population density, territorial interaction and plasma levels of gonadal steroids in spawning male demoiselles *Chromis dispulis* (Pisces: Pomacentridae). Gen. Comp. Endocrinol. 90: 168–176.

Quinn, T.P. 2005. The behavior and ecology of Pacific salmon and trout. University of Washington Press, Seattle. USA.

Quinn, T.P. and C. Groot. 1984. Pacific salmon (*Oncorhynchus*) migrations: orientation vs. random movement. Can. J. Fish. Aqua. Sci. 41: 1319–1324.

Quinn, T.P. and G.M. Tolson. 1986. Evidence of chemically mediated population recognition in coho salmon (*Oncorhynchus kisutch*). Can. J. Zool. 64: 84–87.

Quinn, T.P., B.A. Terjart, and C. Groot. 1989. Migratory orientation and vertical movements of homing adult sockeye salmon, *Oncorhynchus nerka*, in coastal waters. Anim. Behav. 37: 587–599.

Saito, Y., M. Mekuchi, N. Kobayashi, M. Kimura, Y. Aoki, T. Masuda, T. Azuma, M. Fukami, M. Iigo, and T. Yanagisawa. 2011. Molecular cloning, molecular evolution and gene expression of cDNAs encoding thyrotropin-releasing hormone receptor subtypes in a teleost, the sockeye salmon (*Oncorhynchus nerka*). Gen. Comp. Endocrinol. 174: 80–88.

Sato, A., H. Ueda, F. Fukaya, M. Kaeriyama, Y. Zohar, A. Urano, and K. Yamauchi. 1997. Sexual differences in homing profiles and shortening of homing duration by gonadotropin-releasing hormone analog implantation in lacustrine sockeye salmon (*Oncorhynchus nerka*) in Lake Shikotsu. Zool. Sci. 14: 1009–1014.

Sato, K., T. Shoji, and H. Ueda. 2000. Olfactory discriminating ability of lacustrine sockeye and masu salmon in various freshwaters. Zool. Sci. 17: 313–317.

Satou, M., R. Hoshikawa, Y. Sato, and K. Okawa. 2006. An *in vitro* study of long-term potentiation in the carp (*Cyprinus carpio* L.) olfactory bulb. J. Comp. Physiol. A. 192: 135–150.

Scholz, A.T., R.M. Horrall, and A.D. Hasler. 1976. Imprinting to chemical cues: the basis for home stream selection in salmon. Science. 192: 1247–1249.

Shimizu, M., H. Kudo, H. Ueda, A. Hara, M. Shimazaki, and K. Yamauchi. 1993. Identification and immunological properties of an olfactory system-specific protein in kokanee salmon (*Oncorhynchus nerka*). Zool. Sci. 10: 287–294.

Shoji, T., H. Ueda, T. Ohgami, T. Sakamoto, Y. Katsuragi, K Yamauchi, and K. Kurihara. 2000. Amino acids dissolved in stream water as possible homestream odorants for masu salmon. Chem. Sen. 25: 533–540.

Stabell, O.B. 1992. Olfactory control of homing behaviour in salmonids. *In:* T.J. Hara [ed.]. Fish Chemoreception. Chapman and Hall, London. UK pp. 249–270.

Tanaka, H., Y. Naito, N.D. Davis, S. Urawa, H. Ueda, and M. Fukuwaka. 2005. Behavioral thermoregulation of chum salmon during homing migration in coastal waters. Mar. Ecol. Prog. Ser. 291: 307–312.

Thomas, J.D. 1997. The role of dissolved organic matter, particularly free amino acids and humic substances, in freshwater ecosystems. Fresh. Biol. 38: 1–36.

Ueda, H. 1999. Artificial control of salmon homing migration and its application to salmon propagation. Bull. Tohoku Nat. Fish. Res. Inst. 62: 133–139.

Ueda, H. 2004. Recent biotelemetry research on lacustrine salmon homing migration. Mem. Nat. Inst. Polar Res. Spec. Issue. 58: 80–88.

Ueda, H. 2011. Physiological mechanisms of homing migration in Pacific salmon from behavioral to molecular biological approaches. Gen. Comp. Endocrinol. 170: 222–232.

Ueda, H. 2012. Physiological mechanisms of imprinting and homing migration in Pacific salmon *Oncorhynchus* spp. J. Fish Biol. 81: 543–558.

Ueda, H. and K. Yamauchi. 1995. Biochemistry of fish migration. *In:* P.W. Hochachka and T.P. Mommsen [eds.]. Environmental and Ecological Biochemistry, Elsevier, Amsterdam. Netherlands pp. 265–279.

Ueda, H., O. Hiroi, A. Hara, K. Yamauchi, and Y. Nagahama. 1984. Changes in serum concentrations of steroid hormone, thyroxine, and vitellogenin during spawning migration of chum salmon, *Oncorhynchus keta*. Gen. Comp. Endocrinol. 53: 203–211.

Ueda, H., H. Kudo, M. Shimizu, K. Mochida, S. Adachi, and K. Yamauchi. 1993. Immunological similarity between an olfactory system-specific protein and a testicular germ cell protein in kokanee salmon (*Oncorhynchus nerka*). Zool. Sci. 10: 685–690.

Ueda, H., M. Shimizu, H. Kudo, A. Hara, O. Hiroi, M. Kaeriyama, H. Tanaka, H. Kawamura, and K. Yamauchi. 1994. Species-specificity of an olfactory system-specific protein in various species of teleosts. Fish. Sci. 60: 239–240.

Ueda, H., M. Kaeriyama, K. Mukasa, A. Urano, H. Kudo, T. Shoji, Y. Tokumitsu, K. Yamauchi, and K. Kurihara. 1998. Lacustrine sockeye salmon return straight to their natal area from open water using both visual and olfactory cues. Chem. Sen. 23: 207–212.

Ueda, H., J.B.K. Leonard, and Y. Naito. 2000. Physiological biotelemetry research on the homing migration of salmonid fishes. *In:* A. Moore and I. Russell [eds.]. Advances in Fish Telemetry. Crown copyright. Lowestoft. UK pp. 89–97.

Ueda, H., Y. Yamamoto. and H. Hino. 2007. Physiological mechanisms of homing ability in sockeye salmon: from behavior to molecules using a lacustrine model. Am. Fish. Soc. Symp. 54: 5–16.

Ueda, K. 1985. An electrophysiological approach to the olfactory recognition of homestream waters in chum salmon. NOAA Tech. Rep. NMFS. 27: 97–102.

Urano, A., H. Ando, and H. Ueda. 1999. Molecular neuroendocrine basis of spawning migration in salmon. *In:* H.B. Kwon, J.M.P. Joss, and S. Ishii [eds.]. Recent Progress in Molecular and Comparative Endocrinology. Academia Sinica. Taipei. Taiwan pp. 46–56.

Villars, T.A. 1983. Hormones and aggressive behavior in teleost fishes. *In:* B.B. Svare [ed.]. Hormones and Aggressive Behavior. Plenum Press, New York. USA pp. 407–433.

Walker, M.M., C.E. Diebel, C.V. Haugh, P.M. Pankhurst, J.C. Montgomery, and C.R. Green. 1997. Structure and function of the vertebrate magnetic sense. Nature 390: 371–376.

Wickens, A., D. May, and M. Rand-Weaver. 2001. Molecular characterisation of a putative Atlantic salmon (*Salmo salar*) odorant receptor. Comp. Biochem. Physiol. 129B: 653–660.

Wisby, W. J. and A.D. Hasler. 1954. Effect of olfactory occlusion on migrating silver salmon (*Oncorhynchus kisutch*). J. Fish. Res. Board Can. 11: 472–478.

Yamada, H., M. Amano, K. Okuzawa, H. Chiba, and M. Iwata. 2002. Maturational changes in brain contents of salmon GnRH in rainbow trout as measured by a newly developed time-resolved fluoroimmunoassay. Gen. Comp. Endocrinol. 126: 136–143.

Yamamoto, Y. and H. Ueda. 2009. Behavioral responses to natal stream water amino acids in migratory chum salmon. Zool. Sci. 26: 778–782.

Yamamoto, Y., H. Hino, and H. Ueda. 2010. Olfactory imprinting of amino acids in lacustrine sockeye salmon. PLoS ONE. 5: e8633.

Yamauchi, K., N. Koide, S. Adachi, and Y. Nagahama. 1984. Changes in seawater adaptability and blood thyroxine concentrations during smoltification of the masu salmon, *Oncorhynchus masou*, and the amago salmon, *Oncorhynchus rhodurus*. Aquacult. 42: 247–256

Yambe, H., S. Kitamura, M. Kamio, M. Yamada, S. Matsunaga, N. Fusetani, and F. Yamazaki. 2006. L-Kynurenine, an amino acid identified as a sex pheromone in the urine of ovulated female masu salmon. Proc. Nat. Acad. Sci. USA 103: 15370–15374.

Yanagi, S., H. Kudo, Y. Doi, K. Yamauchi, and H. Ueda. 2004. Immunohistochemical demonstration of salmon olfactory glutathione S-transferase class pi (N24) in the olfactory system of lacustrine sockeye salmon during ontogenesis and cell proliferation. Anat. Embryol. 208: 231–238.

Yano, A., M. Ogura, A. Sato, Y. Sakaki, M. Ban, and K. Nagasawa. 1996. Development of ultrasonic telemetry technique for investigating the magnetic sense of salmonids. Fish. Sci. 62: 698–704.

Yano, K. and A. Nakamura. 1992. Observations of the effect of visual and olfactory ablation on the swimming behavior of migrating adult chum salmon, *Oncorhynchus keta*. Jap. J. Ichthyol. 39: 67–83.

Zohar, Y. 1996. New approaches for the manipulation of ovulation and spawning in farmed fish. Bull. Nat. Res. Inst. Aquacult. Suppl. 2: 43–48.

CHAPTER 2

The Behavior and Physiology of Migrating Atlantic Salmon

Andy Moore,* Lucia Privitera and *William D. Riley*

Introduction

The migratory behavior of the anadromous Atlantic salmon (*Salmo salar* L.) has intrigued anglers, scientists and fishery managers for centuries. In his treatise on angling, *The Compleat Angler* published in 1653, Izaak Walton, reports: "Sir Francis Bacon observed, the age of a salmon exceeds not ten years…much of this has been observed by tying a ribbon or some known tape or thread, in the tail of some young salmons, which have been taken in weirs as they have swimmed toward the salt water, and then by taking a part of them again with the known mark at the same place at their return from the sea, which is usually about six months after….which has inclined many to think, that every salmon usually returns to the same river to which it was bred."

Since this publication, knowledge of the life cycle of the salmon has substantially improved and recent innovations in genetic and physiological techniques, together with the rapid development of telemetry technology has permitted both the migratory pathways and the sensory systems underlying these behaviors to be better described both in freshwater and marine environments (Atlantic Salmon Ecology 2011).

Typically, the Atlantic salmon spends the initial stage of its life cycle in freshwater before undergoing a significant transition in terms of morphology, physiology and behavior which permits it to successfully

Cefas, Pakefield Road, Lowestoft, NR33 0HT, Suffolk, United Kingdom.
Email: andy.moore@cefas.co.uk
*Corresponding author

survive the movement into saltwater. Here it spends a number of years feeding before returning to spawn in its river of origin. During its life cycle, the salmon is exposed to a wide range of environmental conditions as well as anthropogenic influences which all combine to challenge its ability to survive and subsequently return to invest in the next generation. At each stage in its migration from the river, through the estuary and into the sea the salmon must negotiate barriers, diffuse and point source pollutants, exploitation, parasites, and predators. In addition, the salmon will also need to cope with changes to its physical environment, as a result of modifications to the river morphology, river flow and discharge and the predicted variations in water temperature. All these factors are likely to combine and interact to various degrees in the future, to regulate populations and pose challenges to all those parties that have a keen interest in the management and conservation of salmon stocks throughout their geographic range.

This chapter endeavors to describe briefly what is known of the migration of the Atlantic salmon and the sensory and physiological mechanisms that control the movements within both the freshwater and marine environments and to discuss how certain changes to their environment, principally from man's activities, can influence the migratory behavior of these iconic fish.

1. The Migration of Atlantic Salmon

Anadromous Atlantic salmon typically spawn in rivers from September to February and eggs hatch the following spring. The timing of hatching and subsequent emergence of the fry varies with latitude and water temperature but the process within each population is timed so that the fry emerge to feed at the most favorable time for survival (Heggberget 1988). After emergence, the fry redistribute themselves downstream and set up feeding territories where as parr, they spend between 1–8 years feeding. Once again, the time spent during this stage of the life cycle depends upon latitude, environmental conditions, growth rate, and genetics.

1.1. Smoltification

After a period of growth in freshwater, the fish prepares for the first major migration stage in their life cycle; the transition from the freshwater to the marine environment. This period is known as smoltification and involves morphological, biochemical, physiological, and behavioral changes that pre-adapt them for life in the sea (Hoar 1988; Høgasen 1998; Thorpe et al. 1998; McCormick et al. 1998; Finstad and Jonsson 2001; Thorstad et al. 2012).

This migration occurs in the spring in all rivers, although the exact timing differs between populations and geographic areas. In general, the smolt migration extends over a three to seven-week period from April to July, with the earliest timing in southern populations (Veselov et al. 1998; Antonsson and Gudjonsson 2002; Stewart et al. 2006; McGinnity et al. 2007; Orell et al. 2007). The majority of individuals belonging to a population, however, may migrate within a relatively short period (1–2 weeks). During smoltification, the salmon parr lose their territorial behavior, show negative rheotaxis and begin schooling. There are also morphological changes which include a slimmer body form and alterations in body coloration (darkened fins, dark back, white belly, and silver sides) that reduce visibility from pelagic predators when in the sea. The final initiation of the smolting process is influenced by photoperiod (increased day length) and water temperature (McCormick and Saunders 1987; McCormick et al. 1998). The physiological changes include modifications to plasma ion concentrations (e.g., chloride Cl– and sodium Na+) and increases in gill Na+K+ATPase activity (Hoar 1988; Boeuf 1993; Strand et al. 2011), thyroid hormones (Iwata 1995; Hutchison and Iwata 1998), growth hormone (GH), cortisol and insulin-like growth factor-I (Hoar 1988; Sakamoto et al. 1995). When smolts are physiologically prepared, an environmental trigger is usually required to initiate downstream migration (McCormick et al. 1998; Riley et al. 2002). The environmental factors cueing downstream migration are mainly water discharge and water temperature. Each of these factors, however, may be of varying importance, and they may stimulate migration in different ways in different populations (Antonsson and Gudjonsson 2002; Carlsen et al. 2004; Davidsen et al. 2005; Jutila et al. 2005). In some rivers, the smolt migration may be initiated solely by changes in water temperature, whereas in other rivers, increased water discharge during the spring spate may be more important (Jonsson and Ruud-Hansen 1985; Hvidsten et al. 1995). Cumulative temperature experienced by the smolts over time may also determine the timing of downstream migration (Zydlewski et al. 2005). In addition, social cues, such as presence of other migrants in the river, may also stimulate migration (Hansen and Jonsson 1985; Hvidsten et al. 1995). It has also been suggested that smolts initiate their downstream migration when environmental cues in the river predict favorable ocean conditions (Hvidsten et al. 2009). However, what is considered most important is that the salmon smolts enter the marine environment during a narrow "window of opportunity" when conditions in the sea are at their optimum in terms of water temperature, prey suitability and availability, which allows survival to be maximized (McCormick et al. 1998).

1.2. Olfactory imprinting

The downstream migration of the smolts is also a period of enhanced olfactory sensitivity (Morin et al. 1989) that allows the fish to imprint to the unique odors of its home river and permits them to return to the same river to spawn after a period in the marine environment (McCormick et al. 1998). The increase in olfactory sensitivity in salmonid smolts has been attributed to changes in thyroid hormone levels (T_4) (Scholz 1980; Hasler and Scholz 1983), and Nevitt and colleagues suggested that the relationship between thyroid hormones and olfactory imprinting might involve differential growth of the peripheral olfactory system (Nevitt et al. 1994). In a study on juvenile coho salmon, Lema and Nevitt demonstrated that even small fluctuations in plasma T_4 are associated with increased rates of proliferation of the olfactory epithelium, establishing a clear link between the thyroid hormone axis and measurable anatomical changes in the peripheral olfactory system (Lema and Nevitt 2004). Interestingly, surges in thyroid hormones have also been implicated in initiating the downstream migratory behavior of salmonid smolts (Iwata 1995; Hutchinson and Iwata 1998). Thyroid hormones may therefore have a dual role in initiating migration and at the same time providing the mechanism by which the smolt imprints to the section of the river where it commences its migration. This would provide the fish with a unique "olfactory snapshot" of its juvenile environment which would enable it to home in with very great precision to the upper reaches of the tributary where it will subsequently spawn. Similarly, the increase in thyroid hormones in smolts as they enter saltwater may also provide a unique olfactory "marker" at the estuary mouth which could be used by the returning adult at the beginning of its spawning migration. Salmon homing behavior, based on their ability to learn environmental cues typical of their home stream results in the development of a distinct population both within a catchment and the associated tributaries. Therefore, each population is genetically and ecologically best adapted to its local environment (Garcia de Leaniz et al. 2007).

1.3. Migratory behavior

Once the fish is physiologically prepared for a life in the marine environment, the smolts begin their downstream migration. Generally, the seaward migration is initiated by an environmental trigger, usually increasing river flow or water temperature.

The downstream migration of smolts within freshwater is an active process; smolts actively swim downstream often within the middle section of the river channel and near the surface where the water velocity is highest and where they are able to avoid areas of reduced flows (Davidsen et al. 2005; Svendsen et al. 2007). Fish tend to migrate at night during the early part of the season when the water temperature is below approximately 12°C. However, towards the end of the season when temperatures are generally higher, smolts migrate during both the day and night as the higher temperature may allow a faster escape response to predators which are more active during the day (Ibbotson et al. 2006; Thorstad et al. 2012). The change in migratory behavior from nocturnal migration to movement during both the day and night is often reflected in a significant seasonal change in the residency time of the smolts within the river. Fish that initiate their migration later in the season spend less time in the river before entering coastal waters. As a result, smolts from a particular river system may migrate through an estuary and enter the sea over a limited number of days (Moore et al. 1995). This period may represent the optimal time or window of opportunity for entry into the marine environment which is critical to their subsequent survival and return as adults (Hansen and Jonsson 1989; Hvidsten et al. 2009). In estuaries smolts use a combination of selective ebb tide transport and active swimming to migrate seawards (Moore et al. 1995). Here the fish also migrate in the upper water column close to the center of the channel where the seaward currents are the highest. This is the most energetically favorable method of moving rapidly through the estuary and into the open sea. Smolt migration through the estuary usually takes place at night, but towards the end of the migrating period and when water temperature is higher, migration also takes place both day and night. The nocturnal pattern of migration in salmon smolts during this period is the result of an endogenous rhythm of swimming activity with the fish moving into the upper water column after dusk, with the light dark cycle acting as a zeitgeber (Moore et al. 1995).

1.4. Autumn smolt migration

Although the main seaward migration of juvenile salmon occurs in the spring, in a number of populations there is also a prior movement of fish during the autumn. Downstream movement of salmon parr during the autumn has been recorded in populations in both North America and Europe (Youngson et al. 1983; Cunjak and Chadwick 1989; Riley et al. 2002), a phenomenon also termed autumn smolt migration. Studies on the River Frome, a chalk stream in southern England, UK have demonstrated that a substantial proportion of the population migrate downstream during the autumn with the peak movement occurring during October and November. This movement can be up to 27% of the spring smolt run (Pinder et al. 2007). The ecological drivers for autumn migrations are unknown (Riley et al. 2008), although a number of mechanisms have been proposed. These include displacement of subordinates by dominant fish (Bjornn 1971; Mason 1976), the requirement for juveniles to migrate to more suitable freshwater habitats (Riddell and Leggett 1981; Huntingford et al. 1992; Riley et al. 2008) or the requirement for mature male parr to locate mature female adults in order to maximize reproductive success. In some cases, the autumn migrations have also been associated with elevated stream discharge (Youngson et al. 1983). Such movements can be composed of predominantly precocious male parr (Buck and Youngson1982) or alternatively of fish of both sexes (Riley et al. 2008). Recently, Hvidsten et al. suggested that smolts initiate their downstream migration when environmental cues in the river predict favorable ocean condition (Hvidsten et al. 2009). Therefore, another potential driver for this movement may be a requirement to move closer to the estuary in order to detect the environmental cues that predict when the ocean conditions are favorable to maximize survival during the transition between the fresh and marine environment. Residency for long periods in the lower river would also allow the fish to imprint to the odors emanating from the whole catchment providing an olfactory signature.

Fig. 1. A rotary screw trap deployed on the River Frome, southern England, to sample the emigrating autumn smolts.

Many autumn migrants, including those that subsequently move to and reside within the tidal reaches of the river during the winter months, are not physiologically adapted to permit permanent, or early, entry into the marine environment (Riley et al. 2008). However, parr that migrate downstream in the autumn do survive and subsequently contribute to the adult stock (Riley et al. 2009). Frequency histograms of seasonal downstream parr movements in the UK suggest a dual peak in the autumn and winter migration, the first occurring in early autumn, the second later during the spawning season for the river system in question (Riley et al. 2002; Pinder et al. 2007; Riley 2007). Although there is often no information from these studies on the sex composition of the migrants, it is speculated that the later migration may involve mature male parr and be related to reproductive activity. These fish are often older than those migrating during the autumn (Riley 2007). The extent to which the timing and relative magnitude of these migrations might vary between rivers or over time is unclear.

1.5. Migration in the marine environment

The relatively recent developments in telemetry methods and technology have expanded the knowledge of the smolt migration within the freshwater environment but also during the early stages of their marine phase and the factors affecting movements and survival (Thorstad et al. 2012). Once smolts leave the estuaries and enter the sea they are called post-smolts. At the beginning of their marine migration, post-smolts are predominantly swimming close to the surface (1–3 m depth) with occasional dives down to 6.5 m (Davidsen et al. 2008). This seems to reflect the tradeoff between the benefit and disadvantages of the different depths and their characteristics such as temperature, salinity, food availability, and predator avoidance (Thorstad et al. 2012). In addition, fish crossing layers of water with different speed and direction may be a mechanism to enable the fish to gather information about the water current direction (Plantalech Manel-la et al. 2009). Movement patterns in coastal waters are complex but the migration is active and seaward but highly variable between individual post-smolts. Salinity seems to be one of the environmental cues that smolts use to move towards the ocean with clear increased swimming speed at higher salinities (Moore et al. 1995; Hedger et al. 2008). Few studies have looked at coastal migration in relation to the day/night cycle and suggestions have been made that swimming speed is higher at night, and possibly linked to migratory movements, while daytime slower movements were attributed to prey detection and predator avoidance (Hedger et al. 2008). In addition, another study indicated that more movements occurred during the night (Dempson et al. 2011).

Once the salmon enters the marine environment, they undertake extensive migrations within the north Atlantic Ocean to their feeding grounds. The temporal and spatial patterns of migration in the sea of salmon from different river systems are not well understood but fish are known to be distributed over a wide area which includes the Norwegian Sea, Labrador Sea, Barent Sea, North Greenland Sea and north of the Faroe Islands (Thorstad et al. 2011). The mechanisms and cues that the salmon use to orientate within the marine environment are still poorly understood.

Sand and Karlsen proposed that the salmon used infrasound patterns within the ocean (Sand and Karlsen 2000). Moore et al. identified biogenic magnetite crystals associated with the lateral line of the salmon which were of suitable shape and size to form a compass system that would allow the fish to orientate over long distances (Moore et al. 1990). However, the mechanisms and sensory systems that allow the salmon as well as other migratory species to navigate over vast distances continue to intrigue fish biologists.

1.6. Spawning migration of adult salmon

Salmon subsequently mature sexually within the marine environment and return to their natal river to spawn. The age at which they mature depends on genetics and the growing conditions in the sea, but the proximate factors initiating the homeward migration are poorly understood (Hansen and Quinn 1998). Orientation between the marine feeding grounds and the coastal zone are considered to be controlled by the same sensory mechanisms as the outward journey. However, it is widely accepted that the final phase of the migration, the movement within the immediate coastal zone is primarily governed by the olfactory discrimination of the home-stream waters. Nordeng proposed through his pheromone hypothesis that fish responded to population specific odors emanating from seaward migrating smolts (Nordeng 1977). However, studies have shown that smolts can return to their home river in the absence of seaward migrating smolts. Solomon suggested that salmon detect the population specific odors from juveniles residing in the upper tributaries (Solomon 1973). This would allow the returning adults, having been to sea for one or more years, an indication of a viable population within the river and that conditions are still suitable for spawning and subsequent juvenile growth.

Salmon return to the rivers many months before they spawn and this timing varies between populations. Fish enter the rivers when the environmental conditions are favorable and the movement through the estuary is completed within a few hours (Solomon and Sambrook 2004; Thorstad et al. 1998). Water discharge is the main proximate factor stimulating salmon to enter the river although other factors such as water temperature, tidal cycle and water quality may all be important (Jonsson and Jonsson 2009; Moore et al. in press). The movement of salmon in freshwater typically includes a quiescent period where fish reside for long periods within pools or below obstruction (Thorstad et al. 2008). The final migration to the spawning grounds occurs typically just prior to reproduction and is dependent upon the reproductive physiology of the fish and proximate cues including water discharge, temperature, and olfactory cues (Moore et al. in press). The olfactory cues that the adults use to home to the spawning grounds are described in great detail in the Chapter 1 in this volume by Ueda. In Atlantic salmon, there is evidence that during this period there is an increase in olfactory sensitivity to certain reproductive cues that are important in synchronizing reproduction between the male and female fish. The olfactory epithelium of the male salmon becomes extremely sensitive to testosterone, which is thought to be released by the mature female and which subsequently attracts the male to the female prior to spawning (Moore and Scott 1991). The ovulated female is then known to release one or more priming pheromones (F-series prostaglandins), which are again detected by the olfactory system of the male and which results in the elevation of plasma sex steroids and an increase in the levels of milt (Moore and Waring 1996a). This increase in olfactory sensitivity would appear to be similar to that which occurred during the imprinting stage when the smolt migrated seaward and may be an important mechanism in the final migration of the males to the spawning grounds, when the females have finally selected a site to excavate for egg deposition. The olfactory cues, together with visual and acoustic cues, are subsequently used in complex reproductive behavior strategies to ensure successful spawning and investment in the next generation (Fleming and Einum 2011).

Atlantic salmon often survive spawning and return to the ocean as kelts. Kelts either return immediately to sea or remain in the river for a number of months (Webb et al. 2007). Kelts generally migrate to sea during the spring or summer after spawning (Niemelä et al. 2000). Whether kelts can be considered to be similar to smolts in terms of their physiology and behavior is not properly understood, but their high survival and rapid movement through the coastal zone may indicate that unlike smolts, the timing of the sea entry is not critical despite their energetic depletion during spawning. The subsequent behavior in the marine environment during this phase of the life cycle is little known although it is thought that kelts from different river systems and population may return to their previous feeding grounds.

2. Factors Affecting the Migration of Atlantic Salmon

The migratory nature of the salmon inevitably exposes it to a wide range of environmental and habitat changes as it moves from freshwater to the marine environment. In particular, the fish must migrate through the estuary at least twice during its life cycle, a transition through one of the most impacted ecosystems in the aquatic environment. As well as the diverse changes in the natural environment, the fish also encounters many anthropogenic modifications to the freshwater and marine environments which will have variable effects at different stages in its life cycle. In terms of migration, any factors that may regulate or inhibit the physiological processes controlling migration or modify the proximate cues and the detection of these cues by the salmon may have a significant impact at the individual and population level. In particular, smoltification can be considered a sensitive stage in the salmon's life history, where a number of factors can modify the timing of downstream migration, imprinting to the home river and successful adaptation to a life in the marine environment. In addition, factors that may have a direct impact on the olfactory system of the adults may result in the inability to detect the home river, straying of fish to other river systems and interference with the olfactory mediated spawning behavior of the adults. There are numerous factors that may pose significant challenges to the salmon but some of the major causes of concern are climate change, pollution, and barriers to migration.

2.1. Climate change

There have been observable changes to the climate in many regions supporting Atlantic salmon. The projected increases in temperature are in the region of 1.1–6.4°C (Solomon et al. 2007) with changes to precipitation patterns and the occurrence of more extreme weather events. These changes are likely to vary geographically but are likely to impact salmon both in freshwater and the marine environment (Todd et al. 2011). The timing of the seaward migration of smolts is often initiated by increasing river discharge and/or water temperature. Reduced river flows are likely to inhibit or delay the emigration of smolts and their entry into coastal waters, in particular in the more southern populations (Jonsson and Jonsson 2009). Any delay in downstream migration may result in reduced survival within the sea and lower numbers of returning adults. Hansen and Jonsson demonstrated that smolts not entering the marine environment during the optimal period for emigration had lower return rates than fish entering during what is considered to be the window of opportunity (Hansen and Jonsson 1989). In regions where there is a marked decrease in the spring river flows, the smolt run could become more dispersed over time and the shoals of fish become smaller. In both instances, this is likely to result in an increase in predation in freshwater, the estuary, and coastal zone. Elevated water temperatures during the freshwater stage of the smolt migration can often lead to the loss of hypo-osmoregulatory capacity; the fish undergo a partial de-smoltification process which reduces survival if they migrate into saline conditions (Duston et al. 1991; Stefansson et al. 1998; McCormick et al. 2000; Handeland et al. 2004). Therefore, there is concern that any delay to the movement of fish into the sea coupled with exposure to elevated temperatures will have a significant impact on the survival of the smolts and the numbers of returning fish. This is of particular concern around structures such as weirs, hydropower schemes and within the impoundments of estuarine barrages where smolts are known to be delayed for long periods and where the water may become warmer (Moore et al. 1996). In addition, the cumulative effects of small delays to smolt migration, caused by a series of obstacles and depleted reaches in the river resulting from increased abstraction of river water within a catchment may be as important to smolt survival as a single large dam or hydropower scheme.

Reduced flows and increased river/estuary temperatures may also inhibit and delay the movement of adult spawning salmon into the freshwater environment. Increased temperatures in the spring and summer months will reduce the amount of suitable thermal habitat for returning salmon, in particular for the southern river basins and reduce the carrying capacity for spawning adults during the quiescent period in their migration. Increased temperatures are also known to have an impact on the reproductive capacity of female salmon and the subsequent reduction in egg deposition and survival. However, where there are increases in precipitation the subsequent increases in river flow may facilitate upstream spawning migration and assist the movement around obstacles such as weirs and barrages. Increased temperature may also result in earlier migrations in the season, later spawning, younger age at sexual maturity, and increased disease susceptibility and mortality. The impacts of climate change are likely to be highest where there are significant pressures on water resources such as southern England, where increasing abstraction will be coupled with decreases in precipitation (but also at the southern edge of population distribution range). Under these circumstances of increased water temperature and low river levels, the two factors may interact to increase the impact on salmon. For instance, during conditions of reduced flows and higher temperatures, certain contaminants will be concentrated and their toxicological effects enhanced, as well as the potential for an increase in infectious diseases, parasites, and pathogens (Harris et al. 2011).

However, there are major uncertainties regarding the impact of changes in climate within the marine environment. The predictions indicate small gradual rises in sea surface temperature, no significant changes, or even slight cooling in those regions occupied by Atlantic salmon. Changes to sea surface temperature and oceanographic features such as currents may modify the distribution and abundance of key prey items of the post-smolts and adult salmon. A mismatch in prey availability during entry into the marine environment may reduce post-smolt survival and growth. Modifications in sea surface temperatures (SST) may reduce the amount of available thermal habitat required for the suitable growth and development of salmon in the sea. This in turn will affect the maturation of salmon in the sea and may modify the timing of the return migration to the rivers. Changes in oceanographic features such as shelf edge currents may compromise the bioenergetic requirements of the migrating fish and lower survival for both the outward emigrating post-smolts as well as the returning adults. In addition, variations between coastal water and freshwater/estuarine temperatures may reduce survival of emigrating smolts and modify the timing of adult returns and entry into freshwater (Todd et al. 2011).

2.2. Pollution and diffuse contaminants

One of the principal factors regulating salmon populations on both sides of the Atlantic Ocean is considered to be pollution and contaminant loading within the freshwater environment. Atlantic salmon is sensitive to a wide range of contaminants emanating from industry, agriculture, aquaculture, and domestic influences. More recently, more novel contaminants such as pharmaceuticals and personal care products are a cause of concern. The Atlantic salmon may be exposed to a wide range of contaminants throughout their life cycle and these may act in isolation or as mixtures to affect a very wide range of biological processes necessary for survival. Point source pollution incidents may cause extensive local mortalities within a population. Episodic releases of contaminated water from industry and mining activities as well as waste from farming activities (slurries) may all provide immediate evidence that water quality has been compromised through losses to the local aquatic populations. However, there is increasing evidence that many contaminants operate in a more subtle way and that sub-lethal effects on salmon populations, although not immediately evident, may nevertheless have serious long term consequences on salmon populations.

For instance any aquatic contaminant that modifies or inhibits the physiological and behavioral processes involved in the parr-smolt transformation, inhibits the detection and response to the proximate cues involved in initiating smolt migration or disrupts successful spawning and reproduction in the salmon will significantly compromise both the individual and the population. Research in this particular area has focused primarily on the role of agricultural pesticides and their effects during sensitive stages in the life cycle of the salmon (reproduction, embryo development, smoltification and saltwater adaptation). In addition much work has also been concentrated on the role of freshwater acidification in regulating salmon populations (Rosseland and Kroglund 2011).

2.2.1. Pesticides and Atlantic salmon

Pesticides that routinely occur in rivers and streams supporting Atlantic salmon fall into four principle groups (Rosseland and Krugland 2011). These are the organophosphate pesticides, carbamate pesticides, triazine pesticides, and pyrethroid pesticides although there are many others that are routinely used throughout northern Europe and the US that do not fit into these groups. Organophosphate pesticides affect the nervous system by disrupting the enzyme that regulates acetylcholine, a neurotransmitter. An example is diazinon which has been used in the UK as a sheep dip insecticide for the control of sheep ticks (Moore and Waring 1996b). Carbamate pesticides also affect the enzyme that regulates the neurotransmitter acetylcholine. An example is carbofuran, an insecticide used to control insects in a wide variety of field crops (Waring and Moore 1997). Triazine pesticides are herbicides used widely in the control of weeds. Atrazine and simazine are common triazine pesticides and have been shown to have a range of effects on salmon (Moore and Waring 1998; Moore and Lower 2001; Moore et al. 2003; Waring and Moore 2004). Pyrethroid pesticides are also insecticides and in the UK were used to replace the organophosphate sheep dip insecticides. The pyrethroids also target the nervous system, particularly the sodium channels in the nerve membranes. Cypermethrin is a common pyrethroid which has been shown to have effects at different stages in the life cycle of Atlantic salmon (Moore and Waring 2001; Lower and Moore 2003) as well as the sea trout (*Salmo trutta* L.) (Jaensson et al. 2007). The following section focuses on how pesticides may affect the physiology, migratory behavior, and sensory systems of Atlantic salmon particularly during sensitive stages in the life cycle such as smoltification, freshwater migration, and reproduction.

2.2.2. Smoltification and smolt migration

One pesticide that has received a great deal of attention regarding its possible impact upon the salmon and the aquatic environment is atrazine. Atrazine (2-chloro-4-ethylamino-6-isopropylamino-s-triazine) is a water-soluble pre- and post-emergence herbicide for the control of annual and perennial grass and annual broad-leaved weeds. The pesticide's use has been restricted within the European Union but it is still one of the most widely used herbicides in the world (and it is still detected in most rivers in the UK).

Salmon smolts exposed to low environmental levels of atrazine whilst still in freshwater were physiologically stressed with increased plasma cortisol, increased osmolarity and variations in monovalent ion concentrations (Waring and Moore 2004; Nieves-Puigdoller et al. 2007), although freshwater exposure to atrazine did not result in any mortalities in the fish. However, when fish previously exposed to atrazine were transferred into saltwater there was a reduction in gill Na^+K^+ ATPase activity, which is an indication of the smolts ability to adapt to saline conditions, and subsequent mortalities in the fish within 24 hours (Waring and Moore 2004).

During their transition between freshwater and saltwater, the salmon smolts will be exposed to a range of contaminants that may operate individually, have an additive, or in certain cases, have a synergistic effect on migration and the ability of fish to tolerate and survive within the sea. Mixtures of contaminants may therefore be important in influencing survival of salmon and regulating populations. Two such compounds that have been shown to operate together to effect smolt physiology and hypo-osmoregulatory performance are atrazine and the endocrine disrupting chemical 4-nonylphenol (4-NP). Once again exposure of salmon smolts in freshwater to environmental levels of the estrogenic chemical 4-nonylphenol (4-NP) during the peak migration period had no significant effect on gill Na+K+ATPase activity, plasma vitellogenin (VTG) levels or hypo-osmoregulatory performance as indicated by survival in sea water. However, where smolts were exposed to mixtures of 4-NP and the pesticide atrazine at environmental levels, there were significant differences in the gill Na+K+ATPase activity, plasma Cl- and Na+ and increased mortalities when transferred to sea water (Moore et al. 2003).

The sense of smell plays an important role during the imprinting process in salmon and is the principal sensory system that the animal uses to successfully home to its natal river to spawn. Freshwater contaminants that interfere or modify that process may subsequently cause long term deleterious effects on specific populations. Atrazine is known to reduce the olfactory mediated detection of priming pheromones known to be important in the synchronization of spawning between the male and female salmon (Moore and Waring 1998). Olfactory sensitivity is also reduced in salmon smolts exposed to this pesticide (Moore et al. 2007). Atrazine reduced the ability of the smolts to detect con-specific smolt urine, which is considered to contain the population specific odors involved in kin recognition and homing in adults.

These studies clearly demonstrate that in terms of the life cycle of the salmon, the freshwater and marine phases cannot be considered in isolation and that conditions within the freshwater environment can have a significant impact on the salmon's subsequent marine survival (Waring and Moore 2004; Russell et al. 2012). So although the smolt experiences the exposure to the contaminant in freshwater, the ecologically relevant response to that chemical only emerges when the fish enters the sea which may misleadingly suggest that the factors regulating salmon populations may occur predominantly in the marine environment. This phenomena is also true of many other contaminants within freshwater with acidification also affecting survival of smolts once they have migrated into the marine environment (Rosseland and Kroglund 2011; Thorstad et al. in press).

Exposure to the pesticide atrazine has been shown to not only modify the physiological and sensory processes involved in smoltification and saltwater adaptation, but to also inhibit the downstream migration of salmon smolts (Moore et al. 2007). Long term exposure to a level of the pesticide, within the range detected in water courses after run-off from agricultural land in the UK, significantly reduced the migratory activity in smolts during a 28 day period that coincided with the peak smolt migration of the local salmon population. Although there was a reduction in the downstream movement of the smolts, there was no impact upon the periodicity of the migratory activity, suggesting that exposure to the pesticide did not delay the initial onset of the nocturnal migration (Moore et al. 2007). However, where there is a period to allow recovery from the exposure to atrazine, the subsequent migratory behavior of the fish may not be significantly affected (Moore et al. 2008). Therefore, the period between exposure in fresh water and entry into the sea may be critical in terms of whether migration and survival in the marine environment is compromised. Contaminants that occur within estuaries and which the smolts are exposed to immediately prior to saltwater entry may be more of a concern than those occurring in areas of the freshwater environment, where there is a significant period between exposure and the migration of the smolts into the sea (Moore et al. 2008).

2.2.4. Potential impacts on populations

Over a number of decades, there have been serious declines in Atlantic salmon populations throughout their geographic distribution. The principal drivers causing this decline are considered to be operating in both the freshwater and marine environments. The conditions experienced by juvenile salmon whilst still in freshwater is an important factor and there is evidence that exposure to a wide range of contaminants may reduce the survival of individuals during the transition into the marine environment. However, there is very little evidence that individual chemicals directly regulate salmon at the population level. In many rivers, the salmon are exposed to a very wide suite of chemicals that often interact together with other environmental and man-made conditions to reduce the fitness and survival of individual stocks. Finding the "smoking gun" in terms of the management and conservation of this species may therefore be very difficult. In terms of the pesticide atrazine, one approach has been to examine the relationship between the occurrence of the chemical in specific rivers and the number of returning adults. In the River Avon in southern England, the rod catch of salmon over a number of years has been examined in relation to the mean annual levels of atrazine routinely measured in the river. Firstly, the relationship between the annual rod-catch and river atrazine levels the previous year was examined to look for a link between the impact of the pesticide during the seaward migration of the smolts and

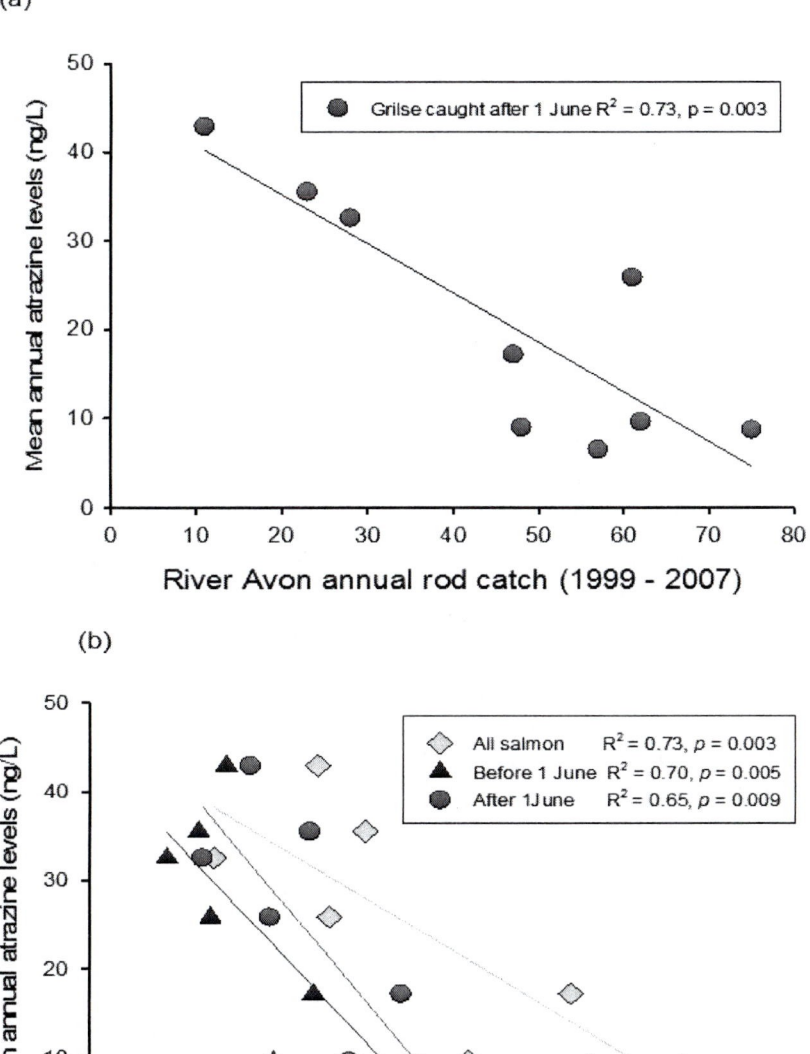

Fig. 2. The relationship between the mean annual levels of atrazine in the River Avon and the annual rod catch of Atlantic salmon. (a) The annual rod catch of grilse returning after one year and the concentration of the pesticide the previous year. (b) The annual rod catch of salmon and the levels of atrazine in the River Avon during the year of return.

the subsequent numbers of returning adults. Secondly, the relationship between the rod-catch and the mean annual atrazine level during the same year was examined to determine whether the pesticide had a direct impact on the number of fish migrating in the river. In both instances the trend was for reduced rod-catch with increasing atrazine levels. These trends can be related to the known toxic effects of atrazine on salmon. It could be suggested that high levels of atrazine during the outward emigration of the smolts reduced marine survival and reduced the numbers of returning adults, whilst atrazine exposure as the adults returned reduced their olfactory abilities to detect home river cues and inhibited migration within the river system. Although such trends may suggest a direct causal link, the deficiencies in using rod-catch as an indication of overall population size and the impact of many other factors operating within river systems requires caution when advising on any management and conservation measures for Atlantic salmon populations.

2.3. Barriers to migration

Implicit to the survival of any migratory animal is the requirement to move unhindered between environments to access the resources required for successful growth and reproduction. The diadromous Atlantic salmon is no exception and one of the major factors regulating populations has been barriers to migration and a loss of connectivity within river systems. Historically, physical barriers such as dams, hydropower schemes, weirs and barrages have been the major challenges to fish movements. However, more recently it has been recognized that more subtle changes to the river system may act as barriers to both the upstream and downstream migration of salmon and these include artificial light.

2.4. Hydropower

Hydropower development has occurred over the last 130 years in a number of countries supporting Atlantic salmon populations (Johnsen et al. 2011). These developments have resulted in serious negative effects for salmon populations in some river systems whilst the effects have been negligible in other rivers. As well as forming physical barriers, the hydropower facilities have a significant impact on river flow and sediment transfer, as well as increasing the mortality of migrating smolts and kelts that pass through the turbines (Skåre et al. 2006). Hydropower dams may impair the timing and success of upstream migrating adults whilst the discharge through the turbine facilities may attract the salmon away from the fish pass and bypass facility. For instance, in the River Umealven in northern Sweden, the discharge from the turbines attracted salmon away from the bypass channel while the increased spill into the bypass channel attracted the salmon to the bypass (Lundquist et al. 2008). In the regulated River Nidelva in southern Norway, salmon migrated to the tunnel outlet of the Rygene power station where they were delayed for up to 71 days (Thorstad et al. 2003).

More recently, there have been proposals to increase the development of low head hydro schemes, particularly within the UK. These small hydroelectric facilities may cause direct mortalities to fish (e.g., where fish encounter turbines) (Calles et al. 2010; Muir et al. 2001) and/or indirect mortalities (e.g., where delays to migration make fish more vulnerable to predation). Of particular concern is the potential cumulative effect of multiple schemes within a river system and the likely impacts on salmon populations which would include delays in fish migration (Holbrook et al. 2009), fish mortality at impoundments and losses of fish spawning and rearing habitats (Hall et al. 2011). The limited understanding of the cumulative effects of these structures compromises the development of effective mitigation measures and to determine the cost-benefit analyses necessary to balance the advantages of renewable energy developments against their potential impacts on fish populations and freshwater communities.

2.5. Estuarine barrages

Estuarine barrages for both tidal power production and for the regeneration of depressed urban areas have been developed on a number of rivers supporting salmon populations. In the UK, amenity barrages have been constructed on the Rivers Tawe and Tees and at Cardiff Bay. More extensive schemes for the production of tidal energy have been proposed on the River Severn in south west England. These structures provide a physical barrier to the movement of salmon even though they include mitigation measures in the form of fish passes to assist the upstream movement of adult fish. The structures also significantly modify the river flow and tidal regime particularly within the impounded river above the barrage. As salmon smolts normally demonstrate a nocturnal, selective ebb tide transport pattern of migration through estuaries (Moore et al. 1995), the removal of the cues necessary for the initiation of migration and rapid transport of fish through the estuary resulted in delays to the movement of smolts through the River Tawe (Moore et al. 1996).

Rapid, nocturnal migration by salmon smolts may reduce the chances of avian predation particularly by cormorants (*Phalacrocorax carbo*) which may prey on smolts in the River Tawe. Predation by birds on salmon smolts may be as high as 70% in some areas (Larsson 1985; Kennedy and Greer 1988), and as the majority of these birds feed visually during the day (Kennedy and Greer 1988), migrating at night may therefore reduce avian predation pressure.

Fig. 3. The River Tees Barrage in north east England, an amenity barrage that provides challenges to the upstream migration of spawning Atlantic salmon. During periods of low flow when fish congregate below the barrage, predation by seals is common.

Color image of this figure appears in the color plate section at the end of the book.

There has also been concern about the potential impact of changes in water quality within impounded estuaries on the survival of salmonid smolts. The modified tidal cycle and retention of river water upstream of barrages may concentrate contaminants and reduce water quality within the impoundment. The delay to smolt migration caused by the structure will also increase the exposure time of fish to any adverse conditions and predation.

Estuarine barrages may also delay the movement of adult salmon from the lower estuary and into freshwater (Russell et al. 1998). In the River Tawe, adult salmon were attracted to the flow of water from the fish pass but did not pass the structure using this facility, however moving over the top of the barrage when the tidal height was sufficient for them to do so (Russell et al. 1998). During this period the fish may be vulnerable to increased predation, particularly by seals (Bendall and Moore 2008).

2.6. Light

There is growing concern regarding the potential impact of increasing levels of artificial light on ecosystems throughout the UK and Northern Europe (Longcore and Rich 2004; Royal Commission on Environmental Pollution 2009; Sutherland et al . 2006). Although terrestrial ecologists have observed that artificial light at night may disrupt migrations, feeding, and other important ecological functions, little is known about the potential role of artificial light in disrupting freshwater ecosystems (Perkin et al. 2011). Artificial light can be considered a form of pollution but where it occurs at specific areas on the migratory route of the salmon, it can also be considered as a potential barrier to the movement of the fish. Photoperiod is generally considered as the primary environmental stimulator regulating the physiological development of salmon smolts (McCormick et al. 1998) and initiating the downstream migration of wild salmon smolts (Hansen and Jonsson 1985; Riley 2007; Riley al. 2002). Recently, Riley et al. clearly demonstrated that artificial street lighting disrupts the diel migratory pattern of wild Atlantic salmon smolts (Riley et al. 2012). Hansen and Jonsson also observed that the speed of descending salmon smolts was reduced under river illumination (Hansen and Jonsson 1985). It is likely that artificial light may also modify or inhibit the upstream movements of adult salmon. This may occur where artificial light is present on existing structures such as fish passes, hydropower schemes, fish counters, and in spawning tributaries situated in areas of high population densities where street lighting is common. Further research is required to assess and quantify how light may affect the migratory stages of the salmon, particularly within river estuaries where port facilities and associated artificial lighting may pose a problem during the transition of salmon between the freshwater and marine environments.

Conclusion

The intriguing nature of the Atlantic salmon's long distance migration will continue to foster a very strong interest in its biology, management, and conservation. The more we understand about all aspects of the behavior of this iconic fish, the better we will be able to reduce the pressures on populations both within the freshwater and marine environment. However, there are still serious challenges to both the fish and those interested in its long term survival. Increasingly, new and novel contaminants are entering the aquatic environment which may have subtle effects on the biology of the fish throughout its life cycle. Increasing concern is now towards pharmaceuticals, personal care products, flame retardants and micro-plastics, all of which may target physiological processes that have long term impacts on the fish ability to survive the transition between fresh and saltwater, find their way home or reproduce. Clearly, certain contaminants operate only during sensitive stages in the life cycle of the salmon and not others. For instance those having an impact on smoltification and downstream migration may not have a toxic effect during reproduction. A mechanistic approach to quantifying the impact of new contaminants on Atlantic salmon is therefore important in determining not only safe and permissible levels within the environment but on the temporal and spatial usage within a river catchment. Freshwater conditions may also regulate survival once the salmon have migrated to sea, but conversely, conditions within the open ocean may also compromise the behavior of returning adults and their abilities to successfully reproduce in freshwater. More emphasis on identifying and quantifying the potential impact of oceanic conditions on fish returning to freshwater is needed to better manage and conserve the species. Climate change poses a serious challenge to the salmon throughout its geographic range and although there may be a limit to the abilities of fisheries managers to control the causes, there is much that can be accomplished in freshwater to mitigate against the effects, particularly in relation to ensuring adequate flows to permit the successful movements of both juvenile and adult fish. Finally, salmon are just one of many fish that undergo extensive migrations and there are lessons to be learnt from other species many of which are covered in other chapters in this book.

Acknowledgments

The authors would like to thank the members of the Salmon and Freshwater Team, Cefas, Lowestoft, for their invaluable contribution to much of the work covered in this chapter. A large proportion of the research was funded by the Department of Environment, Food and Rural Affairs.

References

Antonsson, K. and S. Gudjonsson. 2002. Variability in timing and characteristics of Atlantic salmon smolt in Icelandic rivers. Transactions of the Amer. Fisher. Soc. 131: 643–655.

Atlantic Salmon Ecology. 2011. Aas, Ø. and S. Einum, A. Klemetsen, and J. Skurdal [eds.] Wiley-Blackwell 467 pp.

Bendall, B. and A. Moore. 2008. Temperature-sensing telemetry—possibilities for assessing the feeding ecology of marine mammals and their potential impacts on returning salmonid populations. Fisher. Manag. Ecol. 15: 339–345.

Bjornn, T.C. 1971. Trout and salmon movements in two Idaho streams as related to temperature, stream flow, cover and population density. Transactions of the American Fisheries Society 100: 423–438.

Boeuf, G. 1993. Salmonid smolting: a pre-adaptation to the oceanic environment. In Fish ecophysiology. Rankin, J.C. and F.B. Jensen [eds.]. Chapman and Hall, London. 105–135.

Buck, R. J. G. and A.F. Youngson. 1982. The downstream migration of precociously mature Atlantic salmon, *Salmo salar* L., parr in autumn; its relation to the spawning migration of mature adult fish. J. Fish Biol. 20: 279–288.

Calles, O., I.C. Olsson, C. Comoglio, P.S. Kemp, L. Blunden, M. Schmitz, and L.A. Greenberg. 2010. Size-dependent mortality of migratory silver eels at a hydropower plant, and implications for escapement to the sea. Freshw. Biol. 55: 2167–2180.

Carlsen, K.T., O.K. Berg, B. Finstad, and T.G. Heggberget. 2004. Diel periodicity and environmental influence on the smolt migration of Arctic charr, *Salvelinus alpinus*, Atlantic salmon, *Salmo salar*, and brown trout, *Salmo trutta*, in northern Norway. Environmen. Biol. Fish. 70: 403–413.

Cunjak, R.A. and E.M.P. Chadwick. 1989. Downstream movements and estuarine residence by Atlantic salmon parr (*Salmo salar*). Can. J. Fish. Aquat. Sci. 46: 1466–1471.

Davidsen, J.G., M.A. Svenning, P. Orell, N. Yoccoz, J.B. Dempson, E. Niemelä, A. Klemetsen, A. Lamberg, and J. Erkinaro. 2005. Spatial and temporal migration of wild Atlantic salmon smolts determined from a video camera array in the sub-arctic River Tana. Fish. Res. 74: 210–222.

Davidsen, J.G., N. Plantalech Manel-la, F. Økland, O.H. Diserud, E.B. Thorstad, B. Finstad, R. Sivertsgård, R.S. McKinley, and A.H. Rikardsen. 2008. Changes in swimming depths of Atlantic salmon *Salmo salar* post-smolts relative to light intensity. J. Fish Biol. 73: 1065–1074.

Dempson, J.B., M.J. Robertson, C.J. Pennell, G. Furey, M. Bloom, M. Shears, L.M.N. Ollerhead, K.D. Clarke, R. Hinks, and G.J. Robertson. 2011. Residency time, migration route and survival of Atlantic salmon *Salmo salar* smolts in a Canadian fjord. J. Fish Biol. 78: 1976–1992.

Duston, J., R.L. Saunders, and D.E. Knox. 1991. Effects of increases in freshwater temperature on loss of smolt characteristics in Atlantic salmon (*Salmo salar*). Can. J. Fish. Aquat. Sci. 48: 164–169.

Finstad, B. and N. Jonsson. 2001. Factors influencing the yield of smolt releases in Norway. Nordic J. Freshw. Res. 75: 37–55.

Fleming I.A. and S. Einum. 2011. Reproductive Ecology: A tale of two sexes. Chapter 2 in Atlantic Salmon Ecology. Aas, Ø. and S. Einum, A. Klemetsen, and J. Skurdal [eds.] Wiley-Blackwell. 33–66.

Garcia de Leaniz, C., I.A. Fleming, S. Einum, E. Verspoor, W.C. Jordan, S. Consuegra, N. Aubin-Horth, D. Lajus, B.H. Letcher, A.F. Youngson, J. Webb, L.A. Vøllestad, B. Villanueva, A. Ferguson, and T.P. Quinn. 2007. A critical review of inherited adaptive variation in Atlantic salmon. Biolog. Rev. 82: 173–211.

Hall, C.J., A. Jordaan, and M.G. Frisk. 2011. The historic influence of dams on diadromous fish habitat with a focus on river herring and hydrological longitudinal connectivity. Landsc. Ecol. 26: 95–107.

Handeland, S.O., E. Wilkinson, B. Sveinsbo, S.D. McCormick, and S.O. Stefansson. 2004. Temperature influence on the development and loss of seawater tolerance in two fast-growing strains of Atlantic salmon. Aquacult. 233: 513–529.

Hansen, L. P. and B. Jonsson. 1985. Downstream migration of hatchery-reared smolts of Atlantic salmon (*Salmo salar* L.) in the River Imsa. Aquacult. 45: 237–248.

Hansen, L.P. and B. Jonsson. 1989. Salmon ranching experiments in the river Imsa: effects of timing of Atlantic salmon (*Salmo salar*) smolt migration on survival to adults. Aquacult. 82: 367–373.

Hansen, L.P. and T.P. Quinn. 1998. The marine phase of the Atlantic salmon (*Salmo salar*) life cycle, with comparisons to Pacific salmon. Can. J. Fish. Aquat. Sci. 55 (suppl. 1): 104–118.

Harris, P.D., L. Bachmann, and T.A. Bakke. 2011. The parasites and pathogens of the Atlantic salmon: lessons from *Gyrodactylus salaris*. Chapter 9 in Atlantic Salmon Ecology. Ø. Aas, S. Einum, A. Klemetsen, and J. Skurdal [eds.]. Wiley-Blackwell. 241–252.

Hasler, A.D. and A.T. Scholz. 1983. OlfactoryiImprinting and homing in Salmon. Springer–Verlag, New York.

Hedger, R.D., F. Martin, D. Hatin, F. Caron, F. Whoriskey, and J. Dodson. 2008. Active migration of wild Atlantic salmon *Salmo salar* through a coastal embayment. Mar. Ecol. Prog. Ser. 355: 235–246.

Heggberget, T. 1988. Timing of spawning in Norwegian Atlantic salmon (*Salmo salar*). Canadian Journal of Fisheries and Aquatic Sciences 45: 845–849.

Hoar, W.S. 1988. The physiology of smelting salmonids. Vol XIB. W.S. Hoar, and Randall D.J. [eds.]. Academic Press, New York 275–343.

Holbrook, C.M., J. Zydlewski, D. Gorsky, S.L. Shepard, and M.T. Kinninson. 2009. Movements of prespawn adult Atlantic salmon near hydroelectric dams in the Lower Penobscot River, Maine. North Amer. J. Fish. Manag. 29: 495–505.

Huntingford, F.A., J.E. Thorpe, C. Garcia de Leaniz, and D.W. Hay. 1992. Patterns of growth and smolting in autumn migrants from a Scottish population of Atlantic salmon, *Salmo salar*. J. Fish Biol. 41 (suppl. B): 43–51.

Hutchison, M.J. and M. Iwata. 1998. Effect of thyroxine on the decrease of aggressive behavior of four salmonids during the parr-smolt transformation. Aquacult. 168: 169–175.

Hvidsten, N.A., A.J. Jensen, H. Vivås, Ø. Bakke, and T.G. Heggberget. 1995. Downstream migration of Atlantic salmon smolts in relation to water flow, water temperature, moon phase and social interaction. Nordic J Freshw. Resear. 70: 38–48.

Hvidsten, N.A., A.J. Jensen, A.H. Rikardsen, B. Finstad, J. Aure, S. Stefansson, P. Fiske, and B.O. Johnsen. 2009. Influence of sea temperature and initial marine feeding on survival of Atlantic salmon *Salmo salar* post-smolts from the river Orkla and Hals, Norway. J. Fish Biol. 74: 1532–1548.

Høgåsen, H.R. 1998. Physiological changes associated with the diadromous migration of salmonids. Canadian Special Pubblication in Fisheries and Aquatic Sciences 127: 1078–1081.

Ibbotson, A.T., W.R.C. Beaumont, A. Pinder, S. Welton, and M. Ladle. 2006. Diel migration patterns of Atlantic salmon smolts with particular reference to the absence of crepuscular migration. Ecol. Freshw. Fish 15: 544–551.

Iwata, M. 1995. Downstream migratory behavior of salmonids and its relationship with cortisol and thyroid hormones: a review. Aquaculture 135: 131–139.

Jaensson, A., A.P. Scott, A. Moore, H. Kylin, and K.H. Olsen. 2007. Effects of a pyrethroid pesticide on endocrine responses to female odours and reproductive behavior in male parr of brown trout (*Salmo trutta*). Aquatic Toxicol. 81: 1–9.

Johnsen, B.O., J.V. Arnekleiv, L. Asplin, B.T. Barlaup, T.F. Næsje, B.O. Rosseland, S.J. Saltveit, and A. Tvede. 2011. Hydropower development—Ecological effects. Chapter 14 in Atlantic Salmon Ecology. Ø. Aas, S. Einum, A. Klemetsen, and J. Skurdal [eds.]. Wiley-Blackwell. 351–386.

Jonsson, B. and N. Jonsson. 2009. A review of the likely effects of climate change on anadromous Atlantic salmon *Salmo salar* and brown trout *Salmo trutta*, with particular reference to water temperature and flow. J. Fish Biol. 75: 2381–2447.

Jonsson, B. and J. Ruud-Hansen. 1985. Water temperature as the primary influence on timing of seaward migrations of Atlantic salmon (*Salmo salar*) smolts. Can. J. Fish. Aquatic Sci. 42: 593–595.

Jutila, E., E. Jokikokko, and M. Julkunen. 2005. The smolt run and postsmolt survival of Atlantic salmon, *Salmo salar* L., in relation to early summer water temperatures in the northern Baltic sea. Ecol. Freshw. Fish 14: 69–78.

Kennedy, G.J.A. and J.E. Greer. 1988. Predation by cormorants, *Phalacrocorax carbo* L. on the salmonid populations of an Irish river. Aquacult. Fish. Manag. 19: 159–170.

Larsson, P.O. 1985. Predation on migrating smolt as a regulating factor in Baltic salmon, *Salmo salar* L., populations. Journal of Fish Biology 26: 391–397.

Lema, S.C. and G.A. Nevitt. 2004. Evidence that thyroid hormone induces olfactory cellular proliferation in salmon during a sensitive period for imprinting. J. Exper. Biol. 207: 3317–3327.

Longcore, T. and A. Rich. 2004. Ecological light pollution. Front. Ecol. Environ. 2(4): 191–198.

Lower, N. and A. Moore. 2003. Exposure to insecticides inhibits embryo development and emergence in Atlantic salmon (*Salmo salar* L.). Fish Physiol. Biochem. 28: 431–432.

Lundquist, H., P. Rivinoja, K. Leonardsson, and S. McKinnell. 2008. Upstream passage problems for wild Atlantic salmon (Salmo salar L.) in a regulated river and its effect on the population. Hydrobiol. 602(1): 111–127.

Mason, J.C. 1976. Response of underyearling coho salmon to supplemental feeding in a natural stream. Journal of Wildlife Management 40: 775–788.

McCormick, S.D., L.P. Hansen, T.P Quinn, and R.L. Saunders. 1998. Movement, migration, and smolting of Atlantic salmon (*Salmo salar*). Can. J. Fish. Aquatic Sci. 55 (supp 1): 77–92.

McCormick, S.D., A. Moriyama, and B.T. Bjornsson. 2000. Low temperature limits photoperiod control of smolting in Atlantic salmon through endocrine mechanisms. Amer. J. Physiol. 278: R1352–R1361.

McCormick, S.D. and R.L. Saunders. 1987. Preparatory physiological adaptations for marine life of salmonids: osmoregulation, growth and metabolism. Amer. Fish. Soc. Symp. 1: 211–229.

McGinnity, P., E. de Eyto, T.F. Cross, J. Coughlan, K. Whelan, and A. Ferguson. 2007. Population specific smolt development, migration and maturity schedules in Atlantic salmon in a natural river environment. Aquaculture 273: 257–268.

Moore, A., B. Bendall, J. Barry, C. Waring, N. Crooks, and L. Evans. In press. River temperature and adult anadromous Atlantic salmon, *Salmo salar*, and brown trout, *Salmo trutta*. Fish. Manag. Ecol. Doi: 10.1111/j.1365-2400.2011.00833.x

Moore, A., D. Cotter, G. Rogan, V. Quayle, N. Lower, and L. Privitera. 2008. The impact of a pesticide on the physiology and behavior of hatchery reared salmon smolts during the transition from the freshwater to marine environment. Fish. Manag. Ecol. 15: 385–392.

Moore, A., S.M. Freake, and I.M. Thomas. 1990. Magnetic particles in the lateral line of the Atlantic salmon (*Salmo salar* L.). Phyl. Trans. Royal Soc. B 329: 11–15.

Moore, A. and N. Lower. 2001. The impact of two pesticides on olfactory mediated endocrine function in mature male Atlantic salmon parr. Comp. Biochem. Physiol. B 129: 269–276.

Moore, A., N. Lower, I. Mayer, and L. Greenwood. 2007. The impact of a pesticide on migratory activity and olfactory function in Atlantic salmon (*Salmo salar* L.) smolts. Aquaculture 273: 350–359.

Moore, A., E.C.E. Potter, N.J. Milner, and S. Bamber. 1995. The migratory behavior of wild Atlantic salmon (*Salmo salar*) smolts in the estuary of the river Conwy, North Wales. Can. J. Fish. Aquatic Sci. 52: 1923–1935.

Moore, A. and A.P. Scott. 1991. Testosterone is a potent odorant in precocious male Atlantic salmon (*Salmo salar* L.) parr. Phil. Trans. Royal Soc. Lond. B 332: 241–244.

Moore, A., A.P. Scott, N. Lower, I. Katsiadaki, and L. Greenwood. 2003. The effects of 4-nonylphenol and atrazine on Atlantic salmon (*Salmo salar* L.) smolts. Aquaculture 222: 253–263.

Moore, A., R. Stonehewer, L.T. Kell, M.J. Challiss, M. Ives, I.C. Russell, W.D. Riley, and D.M. Mee. 1996. The movements of emigrating salmonid smolts in relation to the Tawe barrage, Swansea. In Barrages: Engineering Design and Environmental Impacts. N. Burt and J. Watts [eds.]. Chichester: John Wiley and Sons Ltd. pp. 409–417.

Moore, A. and C.P. Waring. 1996a. Electrophysiological and endocrinological evidence that F-series prostaglandins function as priming pheromones in mature male Atlantic salmon parr. J. Exper. Biol. 199: 2307–2316.

Moore, A. and C.P. Waring. 1996b. Sublethal effects of the pesticide Diazinon on olfactory function in mature male Atlantic salmon (*Salmo salar* L.) parr. J. Fish Biol. 48: 758–775.

Moore, A. and C.P. Waring. 1998. Mechanistic effects of a triazine pesticide on reproductive endocrine function in mature male Atlantic salmon parr. Pest. Biochem. Physiol. 62: 41–50.

Moore, A. and C.P. Waring. 2001. The effects of a synthetic pyrethroid pesticide on some aspects of reproduction in Atlantic salmon (*Salmo salar* L.). Aquatic Toxicol. 52(1): 1–12.

Morin, P.-P., J.J. Dodson, and F.Y. Dore. 1989. Cardiac responses to natural odorants as evidence of a sensitive period for olfactory imprinting in young Atlantic salmon, *Salmo salar*. Can. J. Fish. Aquatic Sci. 46: 122–130.

Muir, W.D., S.G. Smith, J.G. Williams, and B.P. Sandford. 2001. Survival of juvenile salmonids passing through bypass systems, turbines, and spillways with and without flow detectors at Snake River dams. North Amer. J. Fish. Manag. 21: 135–146.

Nevitt, G.A., A.H. Dittman, T.P. Quinn, and W. Moody. 1994. Evidence for a peripheral olfactory memory in imprinted salmon. Proc. Nat. Acad. Sci. USA 91: 4288–4292.

Niemelä, E., T.S. Mäkinen, K. Moen, E. Hassinen, J. Erkinaro, M. Länsman, and M. Julkunen. 2000. Age, sex ratio and timing of the catch of kelts and ascending Atlantic salmon in the subarctic River Teno. J. Fish Biol. 56: 974–985.

Nieves-Puigdoller, K., B.T. Björnsson, and S.D. McCormick. 2007. Effects of hexazinone and atrazine on the physiology and endocrinology of smolt development in Atlantic salmon. Aquatic Toxicol. 84(1): 27–37.

Nordeng, H. 1977. A pheromone hypothesis for homeward migration in anadromous salmonids. Oikos 28: 155–159.

Orell, P., J. Erkinaro, M.A. Svenning, J.G. Davidsen, and E. Niemelä. 2007. Synchrony in the downstream migration of smolts and upstream migration of adult Atlantic salmon in the subarctic River Utsjoki. J. Fish Biol. 71: 1735–1750.

Perkin, E.K., F. Hölker, J.S. Richardson, J.P. Sadler, C. Wolter, and K. Tockner. 2011. The influence of artificial light on stream and riparian ecosystems: questions, challenges, and perspectives. Ecosphere 2(11): 1–15.

Plantalech Manel-la, N., E.B. Thorstad, J.G. Davidsen, F. Økland, R. Sivertsgård, R.S. McKinley, and B. Finstad. 2009. Vertical movements of Atlantic salmon post-smolts relative to measures of salinity and water temperature during the first phase of the marine migration. Fish. Manag. Ecol. 16: 147–154.

Pinder, A.C., W.D. Riley, A.T. Ibbotson, and W.R.C. Beaumont. 2007. Evidence for an autumn downstream migration and the subsequent estuarine residence of 0+ juvenile Atlantic salmon, *Salmo salar* L., in England. J. Fish Biol. 71: 260–264.

Riddell, B.E. and W.C. Leggett. 1981. Evidence of an adaptive basis for geographic variation of body morphology, and time of downstream migration of juvenile Atlantic salmon (*Salmo salar*). Can. J. Fish. Aquatic Sci. 38: 308–320.

Riley, W.D. 2007. Seasonal downstream movements of juvenile Atlantic salmon, *Salmo salar* L., with evidence of solitary migration of smolts. Aquaculture 273: 194–199.

Riley, W.D., B. Bendall, M.J. Ives, N.J. Edmonds, and D.L. Maxwell. 2012. Street lighting disrupts the diel migratory pattern of wild Atlantic salmon, *Salmo salar* L., smolts leaving their natal stream. Aquaculture 330–333: 74–81.

Riley, W.D., M.O. Eagle, and S.J. Ives. 2002. The onset of downstream movement of juvenile Atlantic salmon, *Salmo salar* L., in a chalck stream. Fish. Manag. Ecol. 9: 87–94.

Riley, W.D., A.T. Ibbotson, and W.R.C. Beaumont. 2009. Adult returns from Atlantic salmon, *Salmo salar* L., parr autumn migrants. Fish. Manag. Ecol. 16: 75–76.

Riley, W.D., A.T. Ibbotson, N. Lower, A.C. Cook, A. Moore, S. Mizuno, A.C. Pinder, W.R.C. Beaumont, and L. Privitera. 2008. Physiological seawater adaptation in juvenile Atlantic salmon (*Salmo salar*) autumn migrants. Freshw. Biol. 53: 747–755.

Rosseland, B.O. and F. Kroglund. 2011. Lessons from acidification and pesticides. Chapter 15 in Atlantic Salmon Ecology. Ø. Aas, S. Einum, A. Klemetsen, and J. Skurdal [eds.] Wiley-Blackwell. 387–408.

Royal Commission on Environmental Pollution. 2009. Artificial Light in the Environment. The Stationery Office Limited, London 43 pp.

Russell, I.C., M.W. Aprahamian, J. Barry, I.C. Davidson, P. Fiske, A.T. Ibbotson, R.J. Kennedy, J.C. Maclean, A. Moore, J. Otero, E.C.E. Potter, and C.D. Todd. 2012. The influence of the freshwater environment and the biological characteristics of Atlantic salmon smolts on their subsequent marine survival. ICES J. Mar. Sci. 69: 1563–1573.

Russell, I., A. Moore, S. Ives, L.T. Kell, M.J. Ives, and R.O. Stonehewer. 1998. The migratory behavior of juvenile and adult salmonids in relation to an estuarine barrage. Developments in Hydrobiology. Advances in Invertebrates and Fish telemetry. J.-P. Lagardère and M.-L. Bégout Anras, and G. Claireaux [eds]. Kluwer Academic Publishers pp. 321–334.

Sakamoto, T., T. Hirano, S.S. Madsen, R.S. Nishioka, and H.A. Bern. 1995. Insulin-like growth factor I gene expression during the parr–smolt transformation of coho salmon. Zool. Sci. 12: 249–252.

Sand, O. and H.E. Karlsen. 2000. Detection of infrasound and linear acceleration in fishes. Phil. Trans. Royal Soc. B 355: 1295–1298.

Scholz, A.T. 1980. Hormonal regulation of smolt transformation and olfactory imprinting in Coho salmon. Ph.D. Dissertation. University of Wisconsin.

Skåre, P.E., N.A. Hvidsten, T. Forseth, and H.-P. Fjeldstad. 2006. Smoltutvandring forbi Skotfoss kraftverk I Skiensvassdraget ved bygging av et nytt flomkraftverk. NINA rapport 193.

Solomon, D.J. 1973. Evidence for pheromone-influenced homing by migratory Atlantic salmon, *Salmo salar*, L. Nature 224: 231–2.

Solomon, D.J. and H.T. Sambrook. 2004. Effects of hot dry summers on the loss of Atlantic salmon, *Salmo salar*, from estuaries in South West England. Fish. Manag. Ecol. 11: 353–363.

Solomon, S., D. Quin, M. Manning, R.B. Alley, T. Berntsen, N.L. Bindoff, Z. Chen, A. Chidthaisong, J.M. Gregory, G.C. Hegerl, M. Heimann, B. Hewitson, B.J. Hoskins, F. Joos, J. Jouzel, V. Kattsov, U. Lohmann, T. Matsuno, M. Molina, N. Nicholls, J. Overpeck, G. Raga, V. Ramaswamy, J. Ren, M. Rusticucci, R. Somerville, T.F. Stocker, P. Whetton, R.A. Wood, and D. Wratt. 2007. Technical summary. *In:* Climate change 2007: The physical science basis. Contribution of working group I to the fourth assessment report of the intergovernmental panel on climate change (S. Solomon, D. Quin, M. Manning, Z. Chen, M. Marquis, K.B. Averyt, M. Tignor, and H.L. Miller [eds.]. Cambridge university press.

Stefansson, S.O., A.I. Berge and G.S. Gunnarsson. 1998. Changes in seawater tolerance and gill Na$^+$/K$^+$-ATPase activity during desmoltification in Atlantic salmon kept in freshwater at different temperatures. Aquaculture 168: 271–277.

Stewart, D.C., S.J. Middlemas, and A.F. Youngson. 2006. Population structuring in Atlantic salmon (*Salmo salar*): evidence of genetic influence on the timing of smolt migration in sub-catchment stocks. Ecol. Freshw. Fish 15: 552–558.

Strand, J.E.T., J.G. Davidsen, E.H. Jørgensen, and A.H. Rikardsen. 2011. Seaward migrating Atlantic salmon smolts with low levels of gill Na$^+$, K$^+$ -ATPase activity; is sea entry delayed? Environm. Biol. Fish 90: 317–321.

Sutherland, W.J., S. Armstrong-Brown, P.R. Armsworth, T. Brereton, J. Brickland, C.D. Campbell, D.E. Chamberlain, A.I. Cooke, N.K. Dulvy, N.R. Dusic, M. Fitton, R.P. Freckleton, H.C.J. Godfray, N. Grout, H.J. Harvey, C. Hedley, J.J. Hopkins, N.B. Kift, J. Kirby, W.E. Kunin, D.W. MacDonald, B. Marker, M. Naura, A.R. Neale, T. Oliver, D. Osborn, A.S. Pullin, M.E.A. Shardlow, D.A. Showler, P.L. Smith, R.J. Smithers, J.-L. Solandt, J. Spencer, C.J. Spray, C.D. Thomas, J. Thompson, S.E. Webb, D.W. Yalden, and A.R. Watkinson. 2006. The identification of 100 ecological questions of high policy relevance in the UK. J. Appl. Ecol. 43: 617–627.

Svendsen, J.C., A.O. Eskesen, K. Aarestrup, A. Koed, and A.D. Jordan. 2007. Evidence for non-random spatial positioning of migrating smolts (Salmonidae) in a small lowland stream. Freshw. Biol. 52 1147–1158.

Thorpe, J.E., M. Mangel, N.B. Metcalfe, and F.A. Huntingford. 1998. Modelling the proximate basis of salmonid life-history variation, with application to Atlantic salmon, *Salmo salar* L. Evolut. Ecol. 12: 581–599.

Thorstad, E.B., T.G. Heggberge, and F. Økland. 1998. Migratory behavior of adult wild and escaped farmed Atlantic salmon, *Salmo salar* L., before, during and after spawning in a Norwegian river. Aquacult. Res. 29: 419–428.

Thorstad, E.B., I. Uglem, B. Finstad, F. Kroglund, I.E. Einarsdottir, T. Kristensen, O. Diserud, P. Arechavala-Lopez, I. Mayer, A. Moore, R. Nilsen, B.T. Björnsson, and F. Økland. Reduced early marine survival of Atlantic salmon post-smolts and increased physiological stress caused by freshwater exposure to aluminium and moderate acidification. In press.

Thorstad, E.B., F. Whoriskey, A.H. Rikardsen, and K. Aarestrup. 2011. Aquatic nomads: the life and migrations of the Atlantic salmon. Chapter 1 in Atlantic Salmon Ecology. Ø. Aas, S. Einum, A. Klemetsen, and J. Skurdal [eds.]. Wiley-Blackwell. 1–32.

Thorstad, E.B., F. Whoriskey, I. Uglem, A. Moore, A.H. Rikardsen, and B. Finstad. 2012. A critical life stage of the Atlantic salmon *Salmo salar*: behavior and survival during the smolt and initial post-smolt migration. J. Fish Biol. 81: 500–542.

Thorstad, E.B., F. Økland, K. Aarestrup, and T.G. Heggberget. 2008. Factors affecting the within-river spawning migration of Atlantic salmon, with emphasis on human impacts. Rev. Fish Biol. Fish. 18: 345–371.

Thorstad, E.B., F. Økland, F. Kroglund, and N. Jepsen. 2003. Upstream migration of Atlantic salmon at a power station on the river Nidelva, Southern Norway. Fish. Manag. Ecol. 10: 139–146.

Todd, C.D., K.D. Friedland, J.C. MacLean, N. Hazon, and A.J. Jensen. 2011. Getting into hot water? Atlantic salmon responses to climate change in freshwater and marine environments. Chapter 16 in Atlantic Salmon Ecology. Ø. Aas, S. Einum, A. Klemetsen, and J. Skurdal [eds.]. Wiley-Blackwell. 409–443.

Veselov, A.J., M.I. Sysoyeva, and A.G. Potutkin. 1998. The pattern of Atlantic salmon smolt migration in the Varzuga river (White sea basin). Nordic J. Freshw. Res. 74: 65–78.

Waring, C.P. and A. Moore. 1997. Sublethal effects of a carbamate pesticide on pheromonal mediated endocrine function in mature Atlantic salmon (*Salmo salar* L.) parr. Fish Physiol. Biochem. 17: 203–211.

Waring, C.P. and A. Moore. 2004. The effect of atrazine on Atlantic salmon smolts in freshwater and after saltwater transfer. Aquatic Toxicol. 66: 93–104.

Webb, J., E. Verspoor, N. Aubin-Horth, A. Romakkaniemi, and P. Amiro. 2007. The Atlantic salmon. In The Atlantic salmon: genetics, conservation and management. E. Verspoor, L. Stradmeyer, and J. Nielsen [eds.]. Blackwell Publishing Ltd. pp 17–56.

Youngson, A.F., R.J.G. Buck, T.H. Simpson, and D.W. Hay. 1983. The autumn and spring emigrations of juvenile Atlantic salmon L., from the Girnock Burn, Aberdeenshire, Scotland: environmental release of migration. J. Fish Biol. 23: 625–639.

Zydlewski, G. B, A. Haro, and S.D. McCormick. 2005. Evidence for cumulative temperature as an initiating and terminating factor in downstream migratory behavior of Atlantic salmon (*Salmo salar*) smolts. Can. J. Fish. Aquatic Sci. 62: 68–78.

CHAPTER 3

The Onset Mechanisms of the Spawning Migrations of Anguillid Eels

Ryusuke Sudo[a] and Katsumi Tsukamoto[b]

Introduction

Migratory fishes are among the most interesting types of fishes because they are known to travel long distances for purposes of feeding or reproduction. However, the mechanisms by which these fishes begin these remarkable migrations have often difficult to determine. Diadromous fishes such as salmon and anguillid eels are well known for their migrations in the ocean and freshwater, and a variety of other types of fishes also show distinct migration during their life histories.

The onset mechanisms of fish migration have been conceptualized using a three-step model based on field observations and laboratory experiments on amphidromous ayu (Fig. 1) (Tsukamoto et al. 2009). Ayu, *Plecoglossus altivelis*, spawn in freshwater with their larvae drifting to the sea where they grow for an extended period before they enter freshwater for growth and upstream migration (Tsukamoto et al. 1987). The first step for starting their upstream migration is an age or body size threshold, which is reached depending on the hatching date and the growth rate of the individual. The second step is to reach a particular physiological condition in preparing for

Atmosphere and Ocean Research Institute, The University of Tokyo, Kashiwanoha, Kashiwa, Chiba 277-8564, Japan.
Present address:
[a]The Graduate School of Marine Science and Technology, Tokyo University of Marine Science and Technology, Konan, Minato-ku, Tokyo, 108-8477, Japan.
Email: rsudou0@kaiyodai.ac.jp
[b]College of Bioresource Science, Nihon University, Kameino, Fujisawa, Kanagawa 252-0880, Japan.
Email: tsukamoto.katsumi@nihon-u.ac.jp

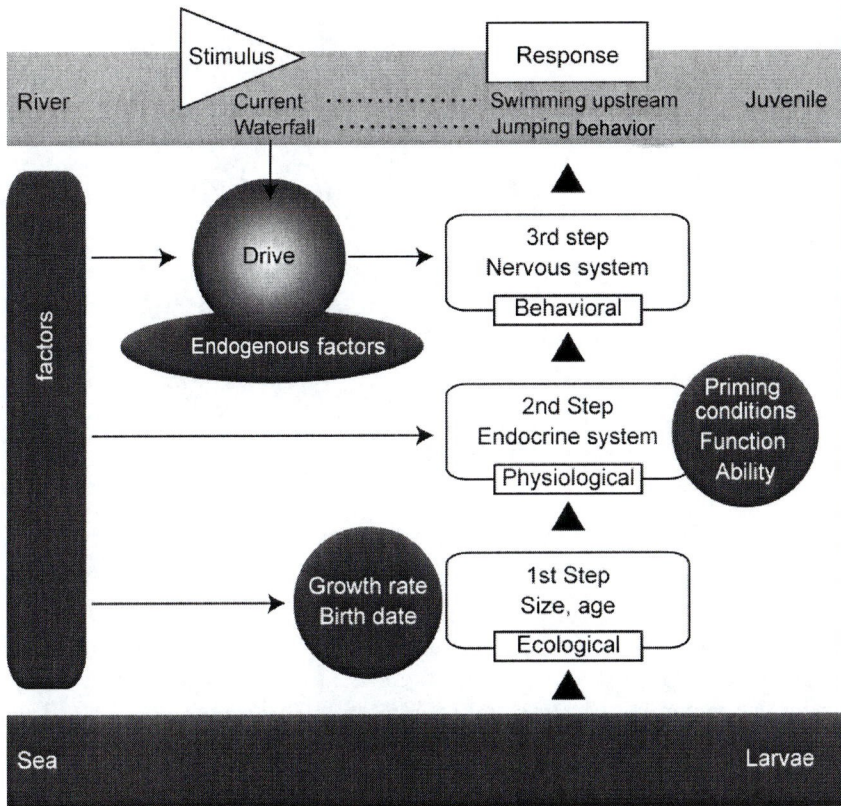

Fig. 1. Three-step model for the releasing mechanism of upstream migration from the sea to freshwater in juvenile ayu. Environmental factors can influence all three steps, and drive can be influenced by endogenous factors such as hunger, temperature, and fish density. The stimuli of current and a waterfall release the upstream migration as the third step through the nervous system. (Modified from Tsukamoto et al. 2009).

migration that includes processes such as smoltification, osmoregulation, and energy accumulation, which are regulated by the endocrine system. As the last step, they need to initiate a behavioral process to release the actual migratory behavior by receiving a trigger from particular exogenous factors and then resultant endogenous ones. Although this three-step model appears to fit well with the upstream migration of ayu, this model has not been evaluated for other diadromous fish such as salmon and eels.

The anguillid eels are catadromous fishes with a complex life cycle that are among the most famous migratory fishes (Fig. 2; Tsukamoto et al. 2011). Their transparent, leaf-like larvae (leptocephali) are transported from the spawning area toward the coastal waters, where they metamorphose into glass eels and probably change their behavior from being pelagic to demersal and begin their inshore migration. Then, they settle in freshwater

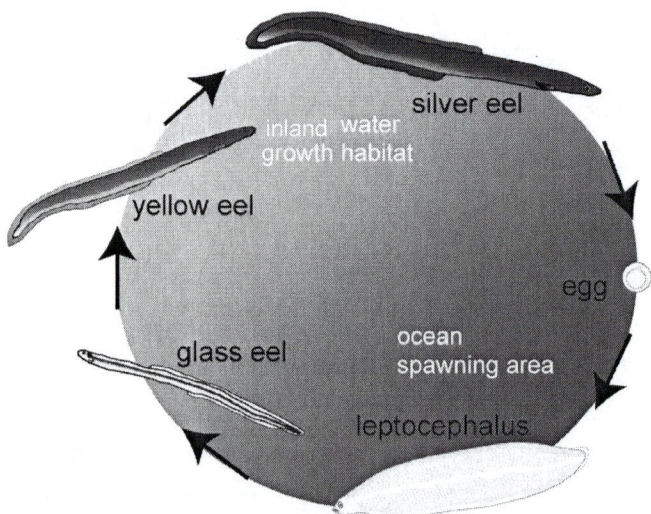

Fig. 2. The catadromous life history of the anguillid eels.

or estuarine habitats as yellow eels. After their growth period, eels change from yellow eels to a migratory phase through a process called silvering. The silver eels then start their spawning migration toward the ocean.

Anguillid eels appear to provide a good model for studying the onset mechanisms of the spawning migrations of diadromous fishes, because the process of the onset of the spawning migration occurs in inland waters where we can observe it more easily than in the ocean. In this chapter, we first review the external and internal morphological changes during silvering, and describe the age and growth of migratory silver eels as a first step. Then, we summarize the environmental factors that may influence the preparation for migration such as priming conditions, silvering and the start of maturation. These environmental factors also act as a trigger for the spawning migration at the third step of the onset mechanism of eel migration. Finally, we describe the endocrinological background of the onset of the spawning migration as the second and third steps of the three-step model, which may be critical to preparing the eels for their long migrations to spawn in the ocean.

1. Silvering of Anguillid Eels

1.1. Change of body coloration

One of the most easily observable changes during silvering is the body color modification, which is most often used for stage determination (Okamura et al. 2007). During the growth phase, yellow eels are usually greenish on the back and yellow-white on the belly. At the onset of the spawning migration, yellow eels change into silver eels with a black pigmented dorsum and a silvery, light colored belly (Fig. 3). This color modification is thought to be the result of increased dorsal melanin and ventral purines (Pankhurst and Lythgoe 1982). Related to this change in body color, the dermis becomes thickened and the total scale area increases (Pankhurst 1982a; Pankhurst and Lythgoe 1982). These changes appear to be adaptations to the pelagic oceanic migration, when they must endure long distance swimming and avoid predators (Tsukamoto 2009).

Fig. 3. Photograph of different stages of the yellow (top) and silver (bottom) Japanese eels, *Anguilla japonica* (Okamura et al. 2007).

Color image of this figure appears in the color plate section at the end of the book.

1.2. Enlargement of eyes

Relative eye size is well known to increase during silvering in anguillid eels. In the European eel, this characteristic is used as one of the indices to distinguish migratory silver eels from sedentary yellow eels (Pankhurst 1982b). It is reported that the number of rods increases and the number of cones appear to decrease since the retinal surface area increases during silvering (Pankhurst 1982b). In addition, their retinal pigments shift from porphyrin and rhodopsin to chrysopsin which is for scotopic vision

(Es-Souni and Ali 1986). These functional modifications of the eyes are thought to be adaptations for the deep-sea environment where light intensity is low.

1.3. Degeneration of the alimentary tract

During hormonally induced sexual maturation, eels exhibit degenerative changes in the alimentary tract (Tesch 2003). There is no quantitative observation in wild silver eels, but numerous authors have found degenerative morphological changes in the stomach and intestines (Pankhurst and Sorensen 1984; Todd 1981; Han et al. 2003a; Tsukamoto 2009). Although the reasons for the degeneration of the digestive tract during silvering are not completely clear, it has been hypothesized that this alteration may facilitate osmoregulation during their diadromous migration and may mostly be a shutdown of an unused system since silver eels stop feeding (Pankhurst and Sorensen 1984). Reduction in size of the intestine appears to be due to atrophy of muscle layers, not the epithelium, which is the important tissue for osmoregulation (Tesch 2003), so this may allow continued water uptake through the gut. In addition, isotopic analysis indicates that eels do not feed during long spawning migration (Chow et al. 2010).

1.4. Development of the swim bladder

Clear modifications of the swim bladder of eels have been found to occur during silvering. The size of the rete mirabile in silver eels was larger than that in yellow eels, and the crystalline guanine deposition in the swim bladder wall of the silver eels was greater than that of the yellow eels (Kleckner 1980a; Kleckner 1980b; Yamada et al. 2001). This results in the gas storage capability in the swim bladder of silver eels being greater than that of yellow eels (Kleckner 1980b). These modifications are considered as an adaptation to enable them to migrate at deeper depths in the ocean than they experienced during the yellow eel phase in rivers and lakes (Tsukamoto 2009). Indeed, recent data recorded by pop-up satellite transmitting tag studies have provided new information about the migratory behavior of silver eels in the ocean. These studies have shown that migrating eels displayed clear patterns of diel vertical migration and reached depths of 600 m–1000 m during daylight hours, while swimming at shallower depths at night (Jellyman and Tsukamoto 2005; Aarestrup et al. 2009; Jellyman and Tsukamoto 2010; Manabe et al. 2011). This explains why they need a stronger swim bladder.

1.5. Maturation of gonads

The gonad weight and gonadosomatic index (GSI, the gonad weight in percent of body weight) increases progressively in female eels during silvering. Gonadal development is highly correlated to other morphological changes during silvering, such as modification of skin color and enlargement of eye size (Svendäng and Wickström 1997; Sbaihi et al. 2001; Han et al. 2003a; Tsukamoto 2009). The increase of GSI can be used as a good criterion to estimate the state of advancement of silvering process in European eels (Durif et al. 2005).

The GSI values of female silver eels appear to be somewhat different among anguillid species apparently as a result of differences in their life histories and migration distances. In the European eel, which has a very long migration, silver eels typically have GSI values of >1.2 but lower than 3.0 (Svendäng and Wickström 1997; Durif et al. 2005). Silver phase Japanese eels (typically > 2.0 and max=4.3) have relatively higher GSI than those of European eels (Sasai et al. 2003; Kotake et al. 2007). This suggests that differences in the rate of maturation of the silver eels at the beginning of their spawning migration between Japanese eel and European eel may be reflecting the differences in their migration distances to their spawning areas (Japanese eel: c. 3,000 km; European eel: c. 6,000 km) (Tsukamoto 2009). This difference was also observed in tropical eels. The GSI values of migrating *A. celebesensis* that migrate only short distances (as short as about 50 km) were over two-fold higher than those of migrating *A. maromorata* whose migratory scale is approximately at least about 1,500 km (Hagihara et al. 2012). Two New Zealand eels, *A. dieffenbachii* and *A. australis*, which are sympatric in New Zealand, also showed differences in GSI values at the beginning of their spawning migration, with the former species having an GSI of 8.1 and the latter 3.5 (Todd 1981); possibly because *A. dieffenbachii* may migrate shorter distances than *A. australis* (Kuroki et al. 2008).

1.6. Age and growth of silver eel

It is known that the age and size of male silver eels are younger and smaller than those of female silver eels. These sexual dimorphisms have been seen in both temperate eels (American eel: Oliveira 1999; European eel: Moriaty 2003; Japanese eel: Yokouchi et al. 2008; *A. australis*: Todd 1980) and some tropical eels (*A. marmorata*: Robinet et al. 2003; *A. reinhardtii*: Walsh et al. 2003). These differences appear to be a uniform characteristic of anguillid eels, and it has been hypothesized that male eels use a time-minimizing strategy and migrate at the earliest possible age, whereas females use 'a size-maximizing strategy' to maximize size and migrate at an optimal body size (Helfman et al. 1987). In male fish, the energy cost of spermatogenesis is thought to be relatively low. Thus, it is sufficient for male eels to reach the minimum size that is needed to complete their migration to reach the spawning site. In female fish, fecundity depends on body size (Wootton 1984) so female eels use 'a size-maximizing strategy' for maximizing fecundity, which is closely related to fitness.

Body size and age at the onset of the spawning migration appear to vary widely, especially in female eels (Table 1). In European eels, age at silvering has been observed to be negatively correlated with growth rate and some of this variation is due to habitat differences (Svendäng et al. 1996). In American eels, it was suggested that the length of female eels increases latitudinally from south to north (Hurley 1972; Facey and LaBar 1981; Helfman et al. 1984; Helfman et al. 1987). Our recent study of Japanese eels revealed that salinity differences influenced the body size and age of female silver eels as well as growth in one drainage system (Sudo et al. in press). These findings indicate that the variability in body size and age of female silver eels is probably a reflection of the variability in habitats and growth conditions of individuals. Oliveira suggested that female American eels may adopt a size-minimizing strategy in less productive habitats, as do the males, but in productive waters they would tend to maximize their size to have greater reproductive potential (Oliviera 1999). The same tendency was found in *A. australis* in New Zealand (Jellyman 2001).

2. Environmental Factors

2.1. *Water temperature*

It is likely that various seasonal cues such as day length or water temperature are important for initiating the onset of the silvering process in anguillid eels. Water temperature decrease is probably an important factor for stimulating the process of silvering that leads to the eels beginning their spawning migration. This is because silver eels were mainly caught during the autumn to winter season in all temperate eels (*A. anguilla*: Vøllestad et al. 1986, Durif and Elie 2008; *A. rostrata*: Haro 1991; *A. japonica*: Okamura et al. 2002; *A. australis* and *A. dieffenbachi*: Todd 1981). In addition, Durif and Elie showed that the onset of spawning migration was affected by low summer water temperatures, which induced an earlier timing of migration (Durif and Elie 2008).

2.2. *Lunar periodicity*

The onset of anguillid migration may also be influenced by lunar periodicity. Many studies have shown that catches of migrating silver European eels increased during the last quarter of the lunar cycle or new moon and

Table 1. Total lengths and ages of silver eels.

Species	Location	Sex	TL (cm)	Age	Reference
A. anguilla	Sweden	Female	60.2-86.8 (n = 324)	12-19 (n = 117)	Svendäng et al. (1996)
	Ireland	Male	28.4-46.0 (n = 152)	10-33 (n = 34)	Poole and Reynolds (1996)
		Female	40.5-92.9 (n = 3283)	8-57 (n = 81)	
	Whole range in	Male	35.7-46.0 (n = 33)	3-15 (n = 34)	Vøllestad 1992
	Europe	Female	40.5-92.9 (n = 32)	4-20 (n = 22)	
A.japonica	China	Male	42.0-59.0 (n = 38)	4-10 (n =32)	Tzeng et al. (2000)
		Female	50.5-70.5 (n = 36)	5-10 (n = 30)	
	Japan	Male	42.6-63.3 (n = 88)	4-16 (n=78)	Yokouchi et al. (2009)
		Female	50.9-80.4 (n = 144)	3-14 (n = 71)	
	Japan	Male	43.7-63.4 (n =31)	4-7 (n=31)	Sudo et al. in press
		Female	57.5-85.6 (n = 61)	4-22 (n = 61)	
A.rostrata	USA	Male	22.8-39.8 (n = 2998)	4-15 (n = 853)	Oliveira (1999)
		Female	40.0-86.7 (n = 137)	6-20 (n = 110)	
	USA	Male	34.6-47.3 (n = 7)	6-18 (n = 65)	Jessop (1987)
		Female	34.6-94.5 (n = 628)	8-43 (n = 628)	
A. australis	New Zealand	Female	65.0-110.7 (n = 116)	9-25 (n = 116)	Jellyman (2001)
	New Zealand	Male	33.8-59.8 (n=12205)	6-24 (n = 1377)	Todd (1980)
		Female	48.3-1024 (n=859)	10-35 (n = 242)	
A. dieffenbachii	New Zealand		48.2-73.5 (n=374)	12-35 (n = 158)	Todd (1980)
			73.7-1560.0 (n=198)	25-60 (n = 20)	

decreased during full moon (Frost 1950; Lowe 1952; Deelder 1954; Todd 1981; Pursianen and Tulonen 1986; Okamura et al. 2002; Miyai et al. 2004; Durif and Elie 2008). Boëtius (1967) found that in laboratory experiments, the number of silver eels escaping from laboratory tanks apparently increased during the last quarter and decreased during the full moon. It is thought that this escape activity may reflect their motivation to migrate.

2.3. Triggering factors

Several possible triggering factors of the spawning migration of anguillid eels are summarized in Table 2. The triggering factors could be divided into hydrological factors that directly affected migratory behavior, and the meteorological factors which were affected by hydrological factors.

Although the lunar cycle may influence eel migrations in the absence of other factors, there are strong indications that weather-related factors may have a stronger influence on the exact timing of migration. It is well known that rainfall is highly correlated with downstream migration of silver eels (Smith and Saunders 1955; Todd 1981; Pursianen and Tulonen 1986; Haro et al. 2002; Boubee et al. 2001). On the other hand, Okamura et al. found no correlation between rainfall and migration in the brackish waters of Mikawa Bay, and suggested that barometric pressure may trigger the spawning migration in coastal areas (Okamura et al. 2002). Strong wind is also thought to be one of the triggering factors of migration (Frost 1950; Cullen and MacCarthy 2003).

Numerous studies reported that river discharge, which was a result of rainfall, has an influence on the migration of eels (Table 2). For example, Winter et al. showed that most eel migration events occurred at distinct moments during a couple of weeks in autumn when the river discharge started to increase (Winter et al. 2006). Rainfall also causes high water level and turbidity. Jens speculated that high water level stimulates eel migration (Jens 1952–1953). Turbidity, which is the result of strong wind and high discharge, may also be related to triggering of migration. Durif et al. showed that downstream eel runs were mostly highly correlated with turbidity and this condition induced higher locomotor activity, which may reflect their migratory drive (Durif et al. 2008). The rise in turbidity can occur both in river and coastal areas, while water levels and discharge does not change. In addition, turbidity influences light intensity and the olfactory organ. These features lead us to think that turbidity may be one of the strongest triggers in silver eel spawning migration.

Table 2. Summary of studies that considered releasing factors that may influence migration behavior in anguillid eels.

Species	Meteorological factor			Hydrological factor			Study
	Rainfall	Barometric pressure	Wind	Discharge	Water level	Turbidity	
A. anguilla			A (+)	A (+)			Frost 1950
A. anguilla				A (+)			Lowe 1952
A. anguilla					A (+)		Jens 1952-1953
A. anguilla				A (+)			Deelder 1954
A. rostrata	A (+)						Smith and Saunders 1955
A. dieffenbachii	A (+)			A (+)			Todd 1981
A. australis	A (+)						Todd 1981
A. anguilla		A (-)		C (+)			Hvidsten 1985
A. anguilla				C (+)			Vøllestad et al. 1986
A. anguilla	A (+)			A (+)			Pursianen and Tulonen 1986
A. anguilla				A (+)			Poole et al. 1990
A. anguilla				A (+)			Jonsson 1991
A. dieffenbachii	A (+)			A (+)			Boubee et al. 2001
A. australis	A (+)						Boubee et al. 2001
A. rostrata	C (+)						Haro et al. 2002
A. japonica		A (-)					Okamura et al. 2002
A. anguilla			A (+)				Cullen and McCarthy 2003
A. anguilla				A (+)		C (+)	Durif et al. 2002
A. anguilla				A (+)			Miyai et al. 2004
A. anguilla				A (+)			Gosset et al. 2005
A. anguilla				A (+)			Winter et al. 2006
A. anguilla				A (+)			Jansen et al. 2007
A. anguilla				A (+)		A (+)	Durif et al. 2008

A: observed association between factor and downstream movement; C: statistical correlation between factor and movement. Sign in parentheses indicates a positive or negative relationship.

3. Endocrinological Mechanisms

3.1 Growth hormone, prolactin, somatolactin

Growth hormone (GH), prolactin (PRL), and somatolactin (SL) are control pleiotropic biological functions in teleosts and are originated from a common ancestral molecule (Rand-Weaver et al. 1992; Vega-Rubín de Celis et al. 2004). GH regulates development and somatic growth (Björnson 1997; Pérez-Sánchez 2000) and is involved as a hypo-osmoregulatory hormone for seawater adaptation in fishes (Sakamoto and McCormick 2006). In contrast, teleost PRL is regarded as a hyper-osmoregulatory hormone for freshwater adaptation (Hirano 1987; Manzon 2002; Sakamoto and McCormick 2006). SL is involved in energy mobilization, stress response, calcium metabolism, acidosis and pigmentation in teleosts (Kawauchi and Sower 2006), although there is little information of its osmoregulatory functions. Furthermore, GH and PRL are involved in regulation of downstream migration in salmonid fishes (Young et al. 1989; Dickhoff et al. 1997; Munakata et al. 2007).

In a recent study, we examined pituitary mRNA expression of these hormones in Japanese eels in relation to salinity difference, silvering, and seasonal change (Fig. 4). Female Japanese eels were collected in the brackish Hamana Lake and its freshwater inlets from July to December and the habitat use histories of the eels were determined using otolith microchemistry (detailed methodology in Sudo et al. 2013). Although GH and PRL have been known to be osmoregulatory hormones, there were no consistent differences in expression levels of these hormones between different salinity habitats. In contrast, SL mRNA expression was higher in eels from freshwater inlets than from the brackish lake. We also showed that PRL mRNA and SL mRNA decreased in the brackish lake and PRL mRNA increased in freshwater inlets from autumn to early winter (Fig. 4).

GH mRNA expression clearly decreased during silvering in our research on the Japanese eel, and this result was consistent with previous research on European eels (Aroua et al. 2005; van Ginneken et al. 2007). PRL mRNA expression did not change during silvering in our study in Hamana Lake system. However, a previous study reported that PRL mRNA levels significantly decreased during silvering (Han et al. 2003b), so the expression of PRL mRNA during silvering continues to remain unclear. SL mRNA expression also did not change clearly during silvering in our study, suggesting that SL may have no relationship to the spawning migration of anguillid eels. Further research on these hormones is needed.

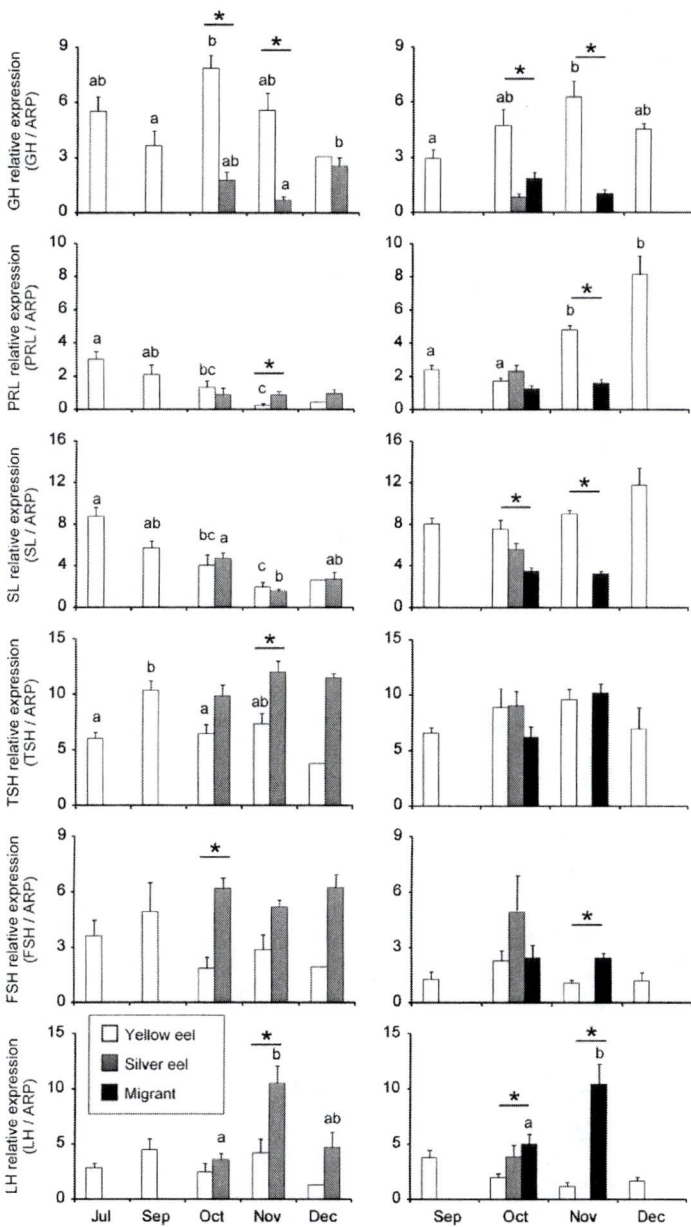

Fig. 4. Monthly changes of mRNA expressions of pituitary hormones for each silvering stage of female eels from both Hamana Lake and its inlet rivers. White bars indicate yellow eels, gray bars indicate silver eels, and black bars indicate migrants. Data are means ± S E. Different letters above the histograms indicate statistically significant differences (p < 0.05). Asterisks indicate significant differences between yellow and silver eels (including migrants).

3.2. Thyroid hormones and thyroid-stimulating hormone

Thyroid hormones, thyroxine (T_4) and triiodothyronine (T_3) are involved in the migration of diadromous fishes. For example, anadromous salmonids require these hormones for the preparatory transformation of smoltification, and are thought to stimulate migratory behavior (Boeuf 1993). In the case of amphidromous ayu, it has been reported that T4 played an important role for the initiation of migration (Tsukamoto 1988). In addition, thyroid hormones also have important roles in controlling the upstream migration of glass eels (Castonguay et al. 1990; Edeline et al. 2005).

Our group showed that there is no significant difference in the plasma levels of thyroid hormones between yellow and silver eels (Fig. 5). This is consistent with other research that showed a moderate increase in T4 and no significant variations in T_3 during silvering (Marchelidon et al. 1999; Aroua et al. 2005; Han et al. 2004). In addition, T_3 treatment did not induce any changes in eye size or the digestive tract of yellow eels (Aroua et al. 2005). Our group and others also measured the pituitary mRNA expression of thyroid-stimulating hormone b subunits (TSHβ), which controls the synthesis of thyroid hormones. Measurement of mRNA expression of TSH showed a non-significant or a weak increase in TSH mRNA between yellow and silver eels (Aroua et al. 2005; Han et al. 2004) (Fig. 4). These results indicated that the thyrotropic axis is not clearly implicated in eel silvering.

3.3. Gonadotropins

As mentioned above, gonad weight sharply increased during silvering in eels. Histological observations revealed the oocytes of silver eels were mainly in the primary yolk globule stage, while those of yellow eels were mainly in early oil drop stages in the Japanese eel (Sudo et al. 2012). The development of gonads is regulated by gonadotropins (GTHs) through the mediation of sex steroids and other hormones produced by the gonads. In anguillid eels, as in other tetrapods, there are two distinct GTHs, follicle-stimulating hormone (FSH) and luteinizing hormone (LH). They are also involved in the control of sex steroids and are related to some aspects of fish behavior (Munakata and Kobayashi 2010).

In European eels, Aroua et al. showed that LH and FSH were differentially expressed during the silvering process, with an increase in FSHβ expression and a late increase of LHβ expression (Aroua et al. 2005). In Japanese eels, Han et al. (observed a simultaneous increase in FSHβ and LHβ mRNA expressions (Han et al. 2003c). We also focused on the variation of mRNA levels of gonadotropins in relation to habitat, silvering, and seasonal change. Our results indicated that the LHβ mRNA expressions were

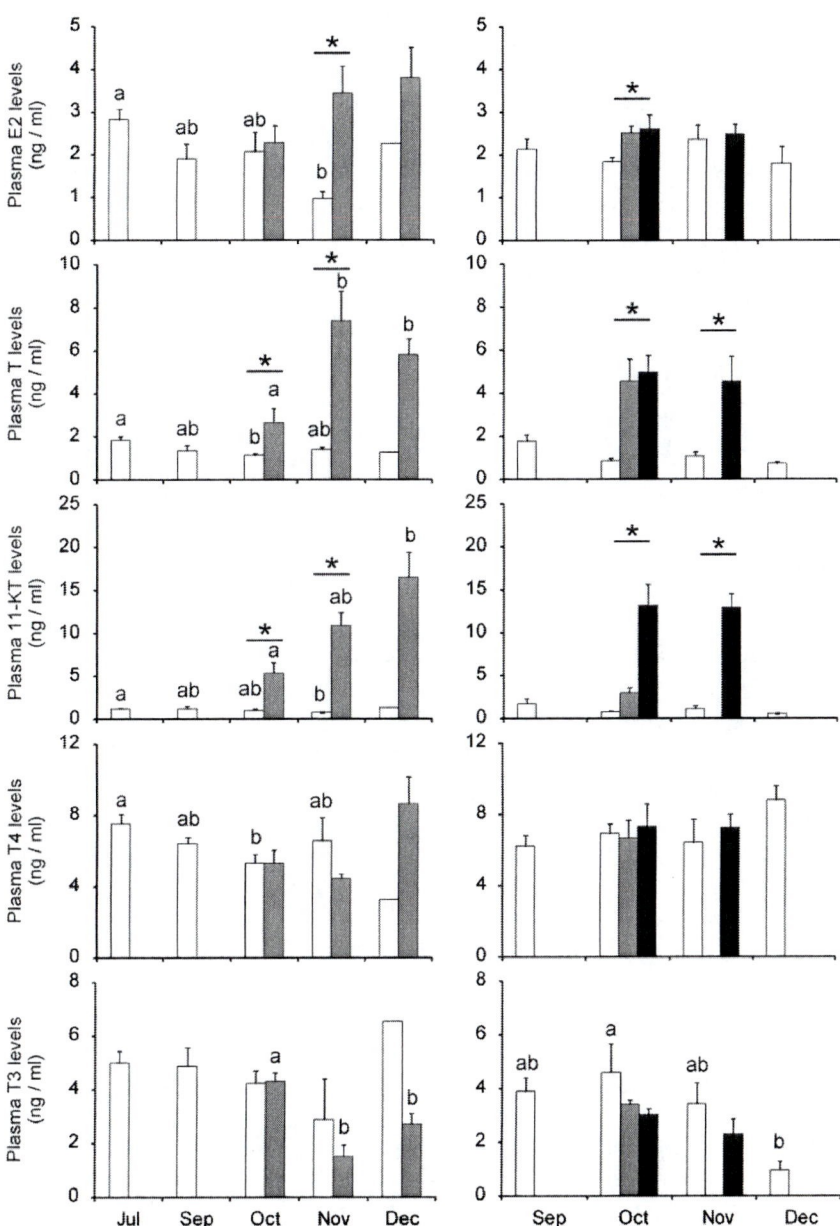

Fig. 5. Monthly changes of steroid hormones (E$_2$, T and 11-KT) and thyroid hormones (T$_4$ and T3) for each silvering stage of female Japanese eels from both the lake and inlets.

significantly higher in silver eels than yellow eels, whereas there were no consistent differences in FSHβ mRNA expression between yellow and silver eels. These expression patterns were similar to both gonadotropin mRNA expression patterns of European eels. We also showed that silver eels not only exhibited high expression levels of FSHβ, but also high LHβ mRNA levels, compared to yellow eels. In addition, it was reported that both FSH and LH receptor mRNA expression significantly increased in experimentally matured eels, which had begun vitellogenesis (Jeng et al. 2007). Together with these findings, it is hypothesized that both FSH and LH are needed for ovarian vitellogenesis, which is one of the features of silvering.

3.4. Sex steroids

Sex steroids are produced in gonads and are regulated by gonadotropins. Among the sex steroids, estrogen (estradiol-17β, E2) is involved in vitellogenesis in ovaries, while androgens (testosterone, T and 11-ketotestosterone, 11-KT) are involved in spermatogenesis in the testis. In addition to their role in fish reproduction, sex steroids are known to be involved in fish growth (Sparks et al. 2003), shifts in body composition (Dasmahapatra et al. 1982), intermediary metabolism (Mommsen and Walsh 1988), osmoregulation (Sangiao-Alvarellos et al. 2006), and migration (Munakata and Kobayashi 2010).

In female eels, increases of E2 during silvering were reported in several species (Lokman et al. 1998; Han et al. 2003d; Aroua et al. 2005), whereas the plasma levels of E2 in male eels were not changed during silvering (Han et al. 2003d; Sudo et al. 2012). In contrast to E2, increases of T during silvering were reported in both female and male eels (Lokman et al. 1998; Lokman and Young 1998; Han et al. 2003d; Aroua et al. 2005; Sudo et al. 2012). Interestingly, 11-KT also drastically increased during silvering in both female and male eels (Lokman et al. 1998; Lokman and Young 1998; Aroua et al. 2005; Sudo et al. 2011a; Sudo et al. 2012), despite 11-KT being traditionally viewed as a male-specific hormone.

3.5. Role of androgen in the onset of spawning migration

From hormone measurements, it is certain that the gonadotropic axis is activated during silvering. Among the various reproductive hormones, androgens are now thought to play an important role in eel silvering process. Olivereau and Olivereau showed that injections of 17α-methyltestosterone into male silver European eel resulted in enlargement of eye diameter, increased skin thickness and darkened head and fins (Olivereau and Olivereau 1985). Similarly, implants of testosterone induced an increase of eye size in male silver eels (Böetius and Larsen 1991). In *A. australis*, it has been reported that an increase in eye diameter, thickening of the skin, and other silvering related changes could be induced by 6-week implants of 11-KT, a non-aromatizable androgen (Rohr et al. 2001). Treatment with T induced a decrease in gut index and an increase in eye index, while E_2 has no effect in the European eel (Aroua et al. 2005). In addition, our research showed that treatment with 11-KT induced the early stage of oocyte growth, enlargement of eye size, degeneration of digestive tract and development of swim bladder in Japanese eels (Fig. 6) (Sudo et al. 2012).

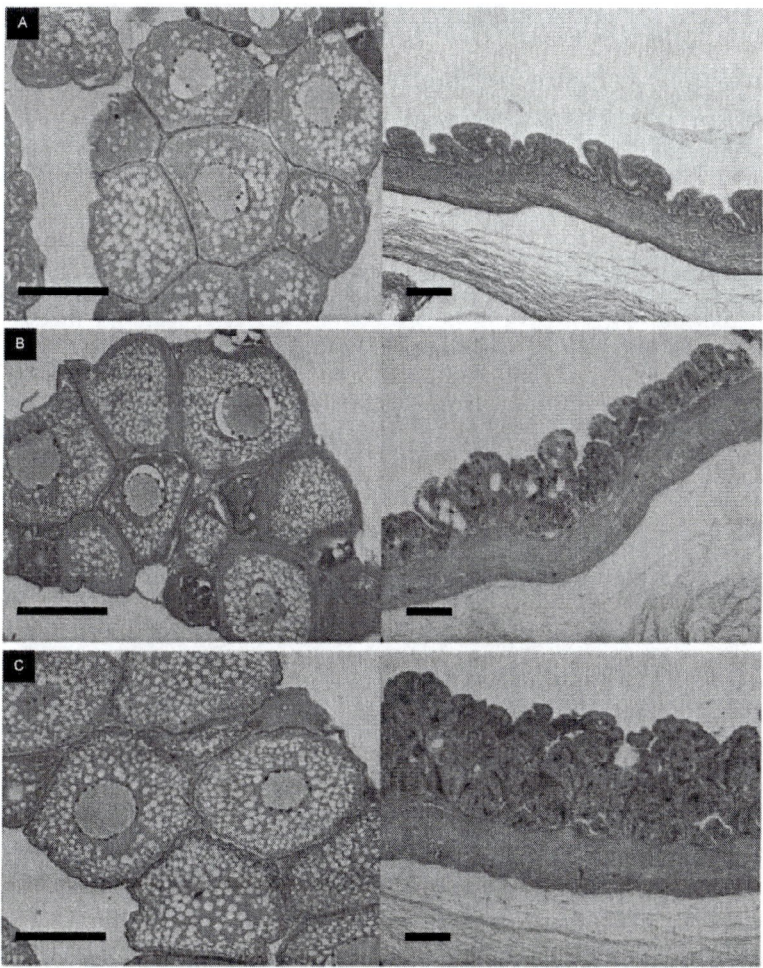

Fig. 6. Effect of 11-KT administration on the ovaries and swim bladder of Japanese eels. Representative micrographs of hematoxylin and eosin stained sections. (A) initial group (B) control group (C) 11-KT group. Right bar = 50 µm. Left bar = 100 µm.

Experimental treatments with androgen clearly revealed that androgens, especially 11-KT accelerates the morphological changes of eel silvering. Although 11-KT did not induce downstream migratory behavior or change of salinity preference, 11-KT treatment induced a higher frequency of movements between fresh water and seawater, which may be related to migratory restlessness (Setiawan et al. 2012). We also demonstrated that 11-KT administration induced increasing locomotor activity of yellow eels in an enclosed tank (Sudo et al. unpublished). In addition, in European eels, androgen was found to stimulate brain dopaminergic systems, which may have an influence on some types of behavior (Weltzien et al. 2006). These results indicated that 11-KT is involved not only in silvering but also in migratory behavior of the onset of spawning migration. Recently, we revealed that gradual water temperature decrease (from 25°C to 15°C) simulating the temperature changes during the autumn migratory season, induced elevation of 11-KT (Fig. 7) (Sudo et al. 2011b). All these findings suggested that 11-KT would be increased by water temperature decrease in autumn, and this would induce silvering and elevate migratory drive. This is in contrast to the downstream migration of salmonids. For instance, precocious masu salmon (*Oncorhynchus masou*) males that had relatively high plasma androgen levels did not show downstream migratory behavior (Aida et al. 1984; Machidori and Katou 1984; Kiso 1995). Androgen administration also inhibited downstream migration in Atlantic salmon (Berglund et al. 1994) and masu salmon (Munakata et al. 2000). It is notable that any effect of androgens on downstream migration should be clearly contrasting in the two types of diadromous fishes because they have

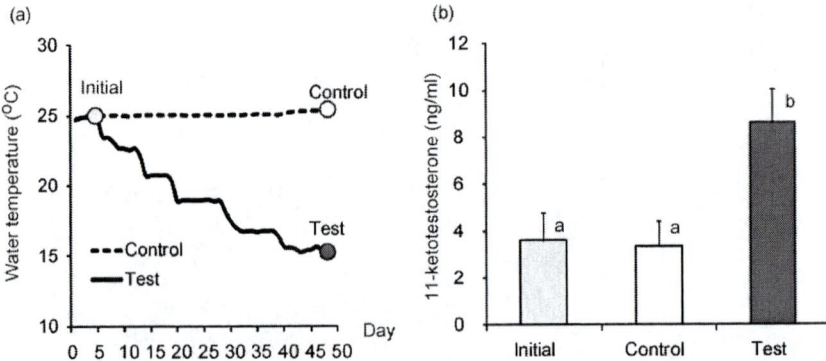

Fig. 7. Effect of temperature decrease on 11-KT levels. (a) Experimental protocol for changes of water temperature during the experiment on influence of water temperature on 11-KT levels with cultured female Japanese eels. Continuous line shows water temperature in the test tank and the broken line shows water temperature in the control tank, with circles indicating the timing of female eel sampling. (b) 11-KT levels (mean ± SE) of initial, control, and test groups. Statistically significant differences between groups are indicated by different letters (p<0.05).

opposite migratory patterns; in catadromous adult anguillid eels, androgens increased during downstream migration, while they inhibit migration in anadromous juvenile masu salmon (Munakata et al. 2000). This may be related to the very different life history stages of eels and salmon, and the differences in motivation between eels, which migrate for spawning, and salmon, which migrate for feeding and growth.

Conclusion

When eels start their spawning migration they change from yellow eels to silver eels, and this process is called silvering. During silvering, eels exhibit various internal and external changes, such as an increase in eye size, degeneration of the digestive tract, and development of the swim bladder, which help the eels pre-adapt to life in the ocean. Silvering also induces changes related to reproductive function, such as significant increase in gonad weight and the modification of oocyte structure. These changes for pre-adaptation to the ocean environment and for reproductive maturation appear to occur synchronously.

The first step for starting their spawning migration is reaching body size threshold, which appears to vary to some degree in female eels of each species (Fig. 8). This variability in body size is probably due to a reflection of the variability in habitats and growth conditions of individuals. The second step is to fulfill a particular physiological condition in preparing for the spawning migration, which is regulated by the endocrine system. During silvering, 11-KT was drastically increased in anguillid eels, and gradual water temperature decrease, simulating the autumn migratory season, was found to induce elevation of 11-KT. The effect of 11-KT administration appeared to induce silvering related changes such as early-stage of oocyte growth, enlargement of eyes, degeneration of digestive tract and development of swim bladder. Thus, it is suggested that in eels that have reached the size or other possible threshold, 11-KT would be increased by water temperature decrease in autumn, and this would induce silvering. As a last step, actual migratory behavior needs a trigger from particular endogenous and exogenous factors. 11-KT may have the potential to elevate the migratory drive that is needed for the onset of spawning migration as endogenous factors. The moon phase may also modulate migratory drive, or influence the timing of migration in the absence of weather-related factors. Finally, factors such as rainfall and strong winds that induce changes in water turbidity may often trigger the onset of migration.

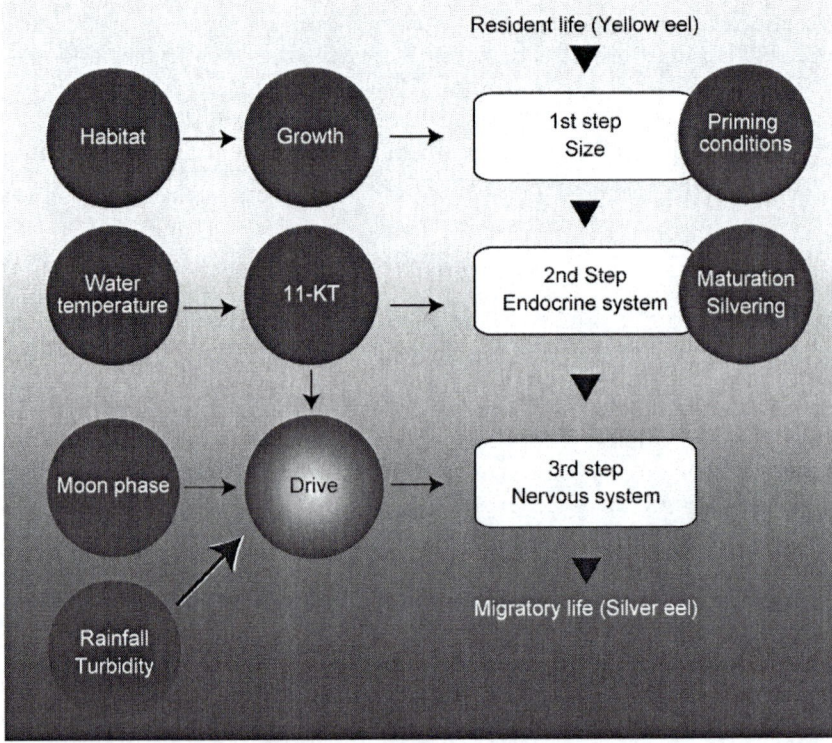

Fig. 8. Three-step model for the onset mechanism of the spawning migration in anguillid eels. Environmental factors can influence all three steps, and drive can be influenced by endogenous factor, 11-KT, and exogenous ones such as moon phase and others. Rainfall and turbidity may release the spawning migration finally in the third phase through the nervous system.

References

Aarestrup, K., F. Okland, M.M. Hansen, D. Righton, P. Gargan, M. Castonguay, L. Bernatchez, P. Howey, H. Sparholt, M.I. Pedersen, and R.S. McKinley. 2009. Oceanic spawning migration of the European eel (*Anguilla anguilla*). Science 325: 1660.

Aida, K., T. Kato, and M. Awaji. 1984. Effects of castration on the smoltification of precocious male masu salmon *Oncorhynchus masou*. Bull. Jpn. Soc. Fish. 50: 565–571.

Aroua, S., M. Schmitz, S. Baloche, B. Vidal, K. Rousseau, and S. Dufour. 2005. Endocrine evidence that silvering, a secondary metamorphosis in the eel, is a pubertal rather than a metamorphic event. Neuroendocrinol. 82: 221–232.

Berglund, I., H. Lundqvist, and H. Fängstam. 1994. Downstream migration of immature salmon (*Salmo salar*) smolts blocked by implantation of androgen 11-ketoandrostenedione. Aquacult. 121: 269–276.

Boëtius, J. 1967. Experimental indication of lunar activity in European silver eels, *Anguilla anguilla*. Medd. Dan. Fisk. Havundes 6: 1–6.

Böetius, I. and L.O. Larsen. 1991. Effects of testosterone on eye size and spermiation in silver eels, *Anguilla anguilla*. Gen. Comp. Endocrinol. 82: 238.

Boeuf, G. 1993. Salmonid smolting: a pre-adaptation to the oceanic environment. *In*: J.C. Rankin and F.B. Jensen [eds.]. Fish Ecophysiology. Chapman & Hall, London UK pp. 105–135.

Boubee, J.A., C.P. Mitchell, B.L. Chisnall, D.W. West, E.J. Bowman, and A. Haro. 2001. Factors regulating the downstream migration of mature eels (*Anguilla* spp.) at Aniwhenua Dam, Bay of Plenty, New Zealand. N. Z. J. Mar. Freshw. Res. 35: 121–134.

Björnsson, B.T. 1997. The biology of salmon growth hormone: from daylight to dominance Fish Physiology and Biochemistry 17: 9–24.

Castonguay, M., J.D. Dutil, C. Audet, and R. Miller. 1990. Locomotor activity and concentration of thyroid hormones in migratory and sedentary juvenile American eels. Trans. Am. Fish Soc. 119: 946–956.

Chow, S., H. Kurogi, S. Katayama, D. Ambe, M. Okazaki, T, Watanabe, T. Ichikawa, M. Kodama, J. Aoyama, A. Shinoda, S. Watanabe, K. Tsukamoto, S. Miyazaki, S. Kimura, Y. Yamada, K. Nomura, H. Tanaka, Y. Kazeto, K. Hata, T. Handa, A. Tawa, and N. Mochioka. 2010. Japanese eel *Anguilla japonica* do not assimilate nutrition during the oceanic spawning migration: evidence from stable isotope analysis. Mar. Ecol. Prog. Ser. 402: 233–238.

Cullen, P. and T.K. McCarthy. 2003. Hydrometric and meteorological factors affecting the seaward migration of silver eels (*Anguilla anguilla*, L.) in the lower River Shannon. Environ. Biol. Fish. 67: 349–357.

Dasmahapatra, A.K. and A.K. Medda 1982. Effect of estradiol dipropionate and testosterone propionate on the glycogen, lipid, and water contents of liver, muscle, and gonad of male and female (vitellogenic and non-vitellogenic) singi fish (*Heteropneustes fossilis* Bloch). Gen. Comp. Endocrinol. 48: 476–484.

Deelder, C.L. 1954. Factors affecting the migration of the silver eel in Dutch inland waters. J. Cons. Perm. Int. Explor. Mer. 20: 177–185.

Dickhoff, W.W., B.R. Beckman, D.A. Larsen, C. Duan, and S. Moriyama. 1997. The role of growth in endocrine regulation of salmon smoltification. Fish. Physiol. Biochem. 17: 231–236.

Durif, C.M.F. and P. Elie. 2008. Predicting downstream migration of silver eels in a large river catchment based on commercial fishery data. Fish. Manag. Ecol. 15: 127–137.

Durif, C., P. Elie, C. Gosset, J. Rives, F. Travade, and D.A. Dixon. 2002. Behavioral study of downstream migrating eels by radio-telemetry at a small hydroelectric power plant. *In:* D. Dixon [eds.]. Biology, management, and protection of Catadromous eels. American Fisheries Society Symposium 33. Bethesda, Maryland, USA pp. 343–356.

Durif, C., S. Dufour, and P. Elie. 2005. The silvering process of *Anguilla anguilla*: a new classification from the yellow resident to the silver migrating stage. J. Fish Biol. 66: 1025–1043.

Durif, C.M.F., F. Travade, J. Rives, P. Elie, and C. Gosset. 2008. Relationship between locomotor activity, environmental factors, and timing of the spawning migration in the European eel, *Anguilla anguilla*. Aquat. Living Resour. 21: 163–170.

Edeline, E., A. Bardonnet, V. Bolliet, S. Dufour, and P. Elie. 2005. Endocrine control of *Anguilla anguilla* glass eel dispersal: Effect of thyroid hormones on locomotor activity and rheotactic behavior. Horm. Behav. 48: 53–63.

Es-Souni, A. and M.A. Ali. 1986. Ultrastructure of the retinal pigmented epithelium of light and dark adapted young pigmented and mature silver eels *Anguilla anguilla*. Zoomorphology 106: 179–184.

Facy, D.E. and G.W. LaBar. 1981. Biology of American eels in Lake Champlain, Vermont. Trans. Am. Fish. Soc. 110: 396–402.

Frost, W.E. 1950. The eel fisheries of the River Bann, Northern Ireland, and observations on the age of silver eels. Cons. Perm. Int. Explor. Mer. 16: 358–393.

Gosset, C., F. Travade, C. Durif, J. Rives, and P. Elie. 2005. Tests of two types of bypass for downstream migration of eels at a small hydroelectric power plant. River Res. Appl. 21: 1095–1105.

Hagihara, S., J. Aoyama, D. Limbong, and K. Tsukamoto. 2012. Morphological and physiological changes of female tropical eels, *Anguilla celebesensis* and *Anguilla marmorata*, in relation to downstream migration. J. Fish Biol. 81: 408–426.

Han, Y.S., I.C. Liao, Y.S. Huang, J.T. He, C.W. Chang, and W.N. Tzeng. 2003a. Synchronous changes of morphology and gonadal development of silvering Japanese eel *Anguilla japonica*. Aquacult. 219: 783–796.

Han, Y.S., J.Y.L. Yu, I.C. Liao, and W.N. Tzeng. 2003b. Salinity preference of silvering Japanese eel *Anguilla japonica*: evidence from pituitary prolactin mRNA levels and otolith Sr: Ca ratios. Mar. Ecol. Prog. Ser. 259: 253–261.

Han, Y.S., I.C. Liao, Y.S. Huang, W.N. Tzeng, and J.Y.L. Yu. 2003c. Profiles of PGH-α, GTH I-β, and GTH II-β mRNA transcript levels at different ovarian stages in the wild female Japanese eel *Anguilla japonica*. Gen. Comp. Endocrinol. 133: 8–16.

Han, Y.S., I.C. Liao, W.N. Tzeng, Y.S. Huang, and J.Y.L. Yu. 2003d. Serum estradiol-17 β and testosterone levels during silvering in wild Japanese eel *Anguilla japonica*. Comp. Biochem. Physiol. B. 136: 913–920.

Han, Y.S., I.C. Liao, W.N. Tzeng, and J.Y.L. Yu. 2004. Cloning of the cDNA for thyroid stimulating hormone β subunit and changes in activity of the pituitary-thyroid axis during silvering of the Japanese eel, *Anguilla japonica*. J. Mol. Endocrinol. 32: 179–194.

Haro, A. 1991. Thermal preference and behavior of Atlantic eels (genus *Anguilla*) in relation to their spawning migration. Environ. Biol. Fish. 31: 171–184.

Haro, A., T. Castro-Santos, L. McLaughlin, K. Whalen, and G. Wipplehauser. 2002. Stimulation of the influence of hydroelectric project operation on mortality of American eels. *In*: D. Dixon [eds.]. Biology, management, and protection of Catadromous eels. American Fisheries Society Symposium 33. Bethesda, Maryland, USA pp. 357–365.

Helfman, G.S., E.L. Bozeman, and B.L. Brothers. 1984. Size, age and sex of American eel in a Geirgia river. Trans. Am. Fish. Soc. 133: 132–141.

Helfman, G.S., D.E. Facey, L.S. Hales, and E.L. Bozeman. 1987. Reproductive ecology of the American eel. *In*: M.J. Dadswell, R.J. Klauda, C.M. Moffitt, R.L. Saunders, R.A. Rulifson, and J.E. Cooper [eds.]. Common strategies of anadromous and catadromous fishes, American Fisheries Society, Symposium 1. Bethesda, Maryland, USA pp. 42–56.

Hirano, T. 1987. Osmoregulatory roles of prolactin and growth hormone in teleosts. Zool. Sci. 4: 1133–1133.

Hurley, D.A. 1972. The American eel (*Anguilla rostrata*) in eastern Ontario. Canadian Journal of Fisheries and Aquatic Sciences 29: 535–543.

Hvidsten, N.A. 1985. Yield of silver eel and factors effecting downstream migration in the stream Imsa Norway. Inst. Freshw. Res. Drottningholm Rep. 62: 75–85.

Jansen, H.M., H.V. Winter, M.C.M. Bruijs, and H.J.G. Polman. 2007. Just go with the flow? Route selection and mortality during downstream migration of silver eels in relation to river discharge. Ices. J. Mar. Sci. 64: 1437–1443.

Jellyman, D. 2001. The influence of growth rate on the size of migrating female eels in Lake Ellesmere, New Zealand. J. Fish Biol. 58: 725–736.

Jellyman, D. and K. Tsukamoto. 2005. Swimming depths of offshore migrating longfin eels *Anguilla dieffenbachii*. Mar. Ecol. Prog. Ser. 286: 261–267.

Jellyman, D. and K. Tsukamoto. 2010. Vertical migrations may control maturation in migrating female *Anguilla dieffenbachii*. Mar. Ecol. Prog. Ser. 404: 241–247.

Jeng, S.R., W.S. Yueh, G.R. Chen, Y.H. Lee, S. Dufour, and C.F. Chang. 2007. Differential expression and regulation of gonadotropins and their receptors in the Japanese eel, *Anguilla japonica*. Gen. Comp. Endocrinol. 154: 161–173.

Jens, G. 1952–1953. Über den lunaren Rhytumus der Blankaalwanderung. Arch. Fischwereiwissenschaft. 4: 94–110.

Jessop, B.M. 1987. Migrating American eels in Nova Scotia. Trans. Am. Fish. Soc. 116: 161–170.

Jonsson, N. 1991. Influence of water flow, water temperature, and light on fish migration in rivers. Nord. J. Freshw. Res. 66: 20–35.

Kawauchi, H. and S.A. Sower. 2006. The dawn and evolution of hormones in the adenohypophysis. Gen. Comp. Endocrinol. 148: 3–14.

Kiso, K. 1995. The life history of masu salmon *Oncorhynchus masou* originated from rivers of the Pacific coast of northern Honshu, Japan. Bull. Natl. Res. Inst. Fish. Sci. 7: 1–188.

Kleckner, R.C. 1980a. Swim bladder wall guanine enhancement related to migratory depth in silver phase *Anguilla rostrata*. Comp. Biochem. Physiol. A. 65: 351–354.

Kleckner, R.C. 1980b. Swim bladder volume maintenance related to initial oceanic migratory depth in silver-phase *Anguilla rostrata*. Science 208: 1481–1482.

Kotake, A., T. Arai, A. Okamura, Y. Yamada, T. Utoh, H.P. Oka, M.J. Miller, and K. Tsukamoto. 2007. Ecological aspect of the Japanese eel, *Anguilla japonica*, collected from coastal areas of Japan. Zool. Sci. 24: 1213–1221.

Kuroki, M., J. Aoyama, M.J. Miller, S. Watanabe, A. Shinoda, D. Jellyman, and K. Tsukamoto. 2008. Distribution and early life history characteristics of anguillid leptocephali in the western South Pacific Ocean. Mar. Fresh. Res. 59: 1035–1047.

Lokman, P.M., G.J. Vermeulen, J.G.D. Lambert, and G. Young. 1998. Gonad histology and plasma steroid profiles in wild New Zealand freshwater eels (*Anguilla dieffenbachii* and *A. australis*) before and at the onset of the natural spawning migration. I. Females. Fish Physiol. Biochem. 19: 325–338.

Lokman, P.M. and G. Young. 1998. Gonad histology and plasma steroid profiles in wild New Zealand freshwater eels (*Anguilla dieffenbachii* and *A. australis*) before and at the onset of the natural spawning migration. II. Males. Fish Physiol. Biochem. 19: 339–347.

Lowe, R.H. 1952. The Influence of Light and Other Factors on the Seaward Migration of the Silver Eel (*Anguilla anguilla* L). J. Anim. Ecol. 21: 275–309.

Machidori, S. and F. Katou. 1984. Spawning populations and marine life of masu salmon (*Oncorhynchus masou*). Int North Pacific Fish Comm. Bull. No. 43, pp. 1–138.

Manabe, R., J. Aoyama, K. Watanabe, M. Kawai, M.J. Miller, and K. Tsukamoto. 2011. First observations of the oceanic migration of the Japanese eel using pop-up tags. Mar. Ecol. Prog. Ser. 437: 229–240.

Manzon, L.A. 2002. The role of prolactin in fish osmoregulation: A review. Gen. Comp. Endocrinol. 125: 291–310.

Marchelidon, J., N. Le Belle, A. Hardy, B. Vidal, M. Sbaihi, E. Buruzawa-Gerard, M. Schmitz, and S. Dufour. 1999. Etude des variations de parametres anatomiques et endocriniens chez l'anguille europeenne (*Anguilla anguilla*) femelle, sedentaire et d'avalaison: application a la caracterisation du stade argente. Bulletin Francais de la Peche Piscicult. 355: 349–368.

Miyai, T., J. Aoyama, S. Sasai, J.G. Inoue, M.J. Miller, and K. Tsukamoto. 2004. Ecological aspects of the downstream migration of introduced European eels in the Uono River, Japan. Environ. Biol. Fish. 71: 105–114.

Mommsen, T.P. and P.J. Walsh. 1988. Vitellogenesis and oocyte assembly. *In*: W.S. Hoar and D.J. Randall [eds.]. Fish physiology. Academic Press, New York, USA pp. 347–406.

Moriaty, C. 2003. The yellow eel. *In*: K. Aida, K. Tsukamoto, and K. Yamauchi [eds.]. Eel Biology. Springer, Tokyo, Japan pp. 89–105.

Munakata, A., M. Amano, K. Ikuta, S. Kitamura, and K. Aida. 2000. Inhibitory effects of testosterone on downstream migratory behavior in masu salmon, *Oncorhynchus masou*. Zool. Sci. 17: 863–870.

Munakata, A., M. Amano, K. Ikuta, S. Kitamura, and K. Aida. 2007. Effects of growth hormone and cortisol on the downstream migratory behavior in masu salmon, *Oncorhynchus masou*. Gen. Comp. Endocrinol. 150: 12–17.

Munakata, A. and M. Kobayashi. 2010. Endocrine control of sexual behavior in teleost fish. Gen. Comp. Endocrinol. 165: 456–468.

Okamura, A., Y. Yamada, S. Tanaka, N. Horie, T. Utoh, N. Mikawa, A. Akazawa, and H.P. Oka. 2002. Atmospheric depression as the final trigger for the seaward migration of the Japanese eel *Anguilla japonica*. Mar. Ecol. Prog. Ser. 234: 281–288.

Okamura, A., Y. Yamada, K. Yokouchi, N. Horie, N. Mikawa, T. Utoh, S. Tanaka, and K. Tsukamoto. 2007. A silvering index for the Japanese eel *Anguilla japonica*. Environ. Biol. Fishes. 80: 77–89.

Oliveira, K. 1999. Life history characteristics and strategies of the American eel, *Anguilla rostrata*. Can J. Fish Aquat Sci. 56: 795–802.

Olivereau, M. and J. Olivereau. 1985. Effects of 17 α-methyltestosterone on the skin and gonads of freshwater male silver eels. General and Comparative Endocrinology 57: 64–71

Pankhurst, N.W. 1982a. Changes in body musculature with sexual maturation in the European eel *Anguilla anguilla* (L.). J. Fish Biol. 21: 549–561.

Pankhurst, N.W. 1982b. Relation of Visual Changes to the onset of sexual maturation in the European eel *Anguilla anguilla* (L.). J. Fish Biol. 21: 127–140.

Pankhurst, N.W. and J.N. Lythgoe. 1982. Structure and color if the tegument of the European eel *Anguilla anguilla* (L). J. Fish Biol. 21: 279–296.

Pankhurst, N.W. and P.W. Sorensen. 1984. Degeneration of the alimentary tract in sexually maturing European eel *Anguilla anguilla* (L.) and American eel *Anguilla rostrata* (LeSeuer). Can. J. Zool. 62: 1143–1148.

Perez-Sanchez, J. 2000. The involvement of growth hormone in growth regulation, energy homeostasis and immune function in the gilthead sea bream (*Sparus aurata*): a short review. Fish Physiol. Biochem. 22: 135–144.

Poole, W.R., J.D. Reynolds, and C. Moriarty. 1990. Observations on the silver eel migrations of the burrishoole river system, Ireland, 1959 to 1988. Internationale Revue Der Gesamten Hydrobiologie. 75: 807–815.

Poole, W.R. and J.D. Reynolds. 1996. Growth rate and age at migration of *Anguilla anguilla* L. J. Fish Biol. 48: 633–642.

Pursiainen, M. and J. Tulonen. 1986. Eel escapement from small forest lakes. Vie et Milieu. 36: 287–290.

Rand-weaver, M., P. Swanson, H. Kawauchi, and W.W. Dickhoff. 1992. Somatolactin, a novel pituitary protein—purification and plasma levels during reproductive maturation of coho salmon. J. Endocrinol. 133: 393–403.

Robinet, T., M. Sbaithi, S. Guyet, B. Mounaix, S. Dufour, and E. Feunteun. 2003. Advanced sexual maturation before marine migration of *Anguilla bicolor bicolor* and *Anguilla marmorata* at Reunion Island. J. Fish Biol. 63: 538–542.

Rohr, D.H., P.M. Lokman, P.S. Davie, and G. Young. 2001. 11-Ketotestosterone induces silvering-related changes in immature female short-finned eels, *Anguilla australis*. Comp. Biochem. Physiol. A 130: 701–714.

Sakamoto, T. and S.D. McCormick. 2006. Prolactin and growth hormone in fish osmoregulation. Gen. Comp. Endocrinol. 147: 24–30.

Sangiao-Alvarellos, S., S. Polakof, F.J. Arjona, A. García-Lopez, M.P. Martín del Rio, G. Martínez-Rodriguez, J.M. Míguez, J.M. Mancera, and J.L. Soengas. 2006. Influence of testosterone administration on osmoregulation and energy metabolism of gilthead sea bream *Sparus auratus*. Gen. Comp. Endocrinol. 149: 30–41.

Sasai, S., J. Aoyama, S. Watanabe, T. Kaneko, M.J. Miller, and K. Tsukamoto. 2001. Occurrence of migrating silver eels Anguilla japonica in the East China Sea. Mar. Ecol. Prog. Ser. 212: 305–310.

Sbaihi, M., M. Fouchereau-Peron, F. Meunier, P. Elie, I. Mayer, E. Burzawa-Gerard, B. Vidal and S. Dufour. 2001. Reproductive biology of the conger eel from the south coast of Brittany, France and comparison with the European eel. J. Fish Biol. 59: 302–318.

Setiawan, A.N., M.J. Wylie, E.L. Forbes, and P.M. Lokman. 2012. The effect of 11-ketotesotrerone on occupation of downstream location and seawater in the New Zealand Shortfinned eel, *Anguilla australis*. Zool. Sci. 29: 1–5.

Smith, M.W. and J.W. Saunders. 1955. The American eel in certain fresh waters of the Maritime Provinces of Canada. J. Fish. Res. Board Can. 12: 238–269.

Sparks, R.T., B. Ron, B.S. Shepherd, S.K Shimoda, G.K. Iwama, and E.G. Grau. 2003. Effects of environmental salinity and 17α-methyltestosterone on growth and oxygen consumption in the tilapia, *Oreochromis mossambicus*. Comp. Biochem. Physiol. B. 136: 657–665.

Sudo, R., H. Suetake, Y. Suzuki, T. Utoh, S. Tanaka, J. Aoyama, and K. Tsukamoto. 2011a. Dynamics of reproductive hormones during downstream migration in females of the Japanese eel, *Anguilla japonica*. Zool. Sci. 28: 180–188.

Sudo, R., R. Tosaka, S. Ijiri, S. Adachi, H. Suetake, Y. Suzuki, N. Horie, S. Tanaka, J. Aoyama, and K. Tsukamoto. 2011b. The effect of temperature decrease on oocyte development, sex steroids and gonadotropin β-subunit mRNA expression levels in female Japanese eels *Anguilla japonica*. Fish. Sci. 77: 575–582.

Sudo, R., R. Tosaka, S. Ijiri, S. Adachi, J. Aoyama, and K. Tsukamoto. 2012. 11-ketotestosterone synchronously induces oocyte development and silvering related changes in the Japanese eel, *Anguilla japonica*. Zool. Sci. 29: 254–259.

Sudo, R., H. Suetake, Y. Suzuki, J. Aoyama, and K. Tsukamoto. 2013. Profiles of mRNA expression for prolactin, growth hormone, and somatolactin in Japanese eel, *Anguilla japonica*: the effect of salinity, silvering and seasonal change Comp. Biochem. Physiol. A 164: 10–16.

Sudo, R., N. Fukuda, J. Aoyama and K. Tsukamoto. Age and body size of Japanese eels, *Anguilla japonica*, at the silver-stage in the Hamana Lake system, Japan. Coast. Mar. Sci. in press.

Svendäng, H.E., Neuman, and H. Wickström. 1996. Maturation patterns in female European eel: age and size at the silver eel stage. J. Fish Biol. 48: 342–351.

Svedäng, H. and H. Wickström. 1997. Low fat contents in female silver eels: indications of insufficient energetic stores for migration and gonadal development. J. Fish Biol. 50: 475–486.

Tesch, F.W. 2003. The eel. London, Blackwell Publishing.

Todd, P.R. 1980. Size and age of migrating New Zealand freshwater eels (*Anguilla* spp.). N. Z. J. Mar. Freshwater Res. 14: 283–293.

Todd, P.R. 1981. Timing of periodicity of migrating New Zealand freshwater eels (*Anguilla* spp.). N. Z. J. Mar. Freshwater Res. 15: 225–235.

Tsukamoto, K. 1987. Switching of size and migratory pattern in successive generations of landlocked Ayu. *In*: M.J. Dadswell, R.J. Klauda, C.M. Moffitt, R.L. Saunders, R.A. Rulifson, and J.E. Cooper [eds.]. Common strategies of anadromous and catadromous fishes. American Fisheries Society Symposium 1. Bethesda, Maryland, USA pp. 492–506.

Tsukamoto, K. 1988. Migratory mechanisms and behavioral characteristics in ayu. *In*: T. Ueno and M. Okiyama [eds.]. Ichthyology currents. Asakurashoten, Tokyo, Japan pp. 100–133.

Tsukamoto, K. 2009. Oceanic migration and spawning of anguillid eels. J. Fish Biol. 74: 1833–1852.

Tsukamoto, K., M.J. Miller, A. Kotake, J. Aoyama, and K. Uchida. 2009. The origin of fish migration: the random escapement hypothesis. *In*: A. Haro, K.L. Smith, R.A. Rulifson, C.M. Moffitt, R.J. Klauda, M.J. Dadswell, R.A. Cunjak, J.E. Cooper, K.L. Beal, and T.S. Avery [eds.]. Challenges for Diadromous Fishes in a Dynamic Global Environment. American Fisheries Society Symposium 69. Bethesda, Maryland, USA pp. 45–61.

Tsukamoto, K., S. Chow, T. Otake, H. Kurogi, N. Mochioka, M.J. Miller, J. Aoyama, S. Kimura, S. Watanabe, T Yoshinaga, A. Shinoda, M. Kuroki, M. Oya, T. Watanabe, K. Hata, S. Ijiri, Y. Kazeto, K. Nomura, and H. Tanaka. 2011. Oceanic spawning ecology of freshwater eels in the western North Pacific. Nat. Commun. 2: 179.

Tzeng, W.N., H.R. Lin, C.H. Wang, and S.N. Xu. 2000. Differences in size and growth rates of male and female migrating Japanese eels in Pearl River, China. J. Fish Biol. 57: 1245–1253

van Ginneken, V., C. Durif, S. Dufour, M. Sbaihi, R. Boot, K. Noorlander, J. Doornbos, A.J. Murk, and G. van den Thillart. 2007. Endocrine profiles during silvering of European eel (*Anguilla anguilla* L.) living in saltwater. Anim. Biol. 57: 453–465.

Vega-Rubín de Celis, S., P. Rojas, P. Gomez-Requeni, A. Albalat, J. Gutierrez, F. Medale, S.J. Kaushik, I. Navarro, and J. Perez-Sanchez. 2004. Nutritional assessment of somatolactin function in gilthead sea bream (*Sparus aurata*): concurrent changes in somatotropic axis and pancreatic hormones. Comp. Biochem. Phys. A. 138: 533–542.

Vøllestad, L.A. 1986. Temperature dependent activity of brackish water yellow eels *Anguilla anguilla*. Aquacult. Fisher. Manag. 17: 201–206.

Vøllestad, L.A. 1992. Geographic variation in age and length at metamorphosis of maturing European eel: environmental effects and phenotypic plasticity. J. Fish Biol. 61: 41–48.

Walsh, C.T., B.C. Pease, and D.J. Booth. 2003. Sexual dimorphism and gonadal development of the Australian longfinned river eel. J. Fish Biol. 63: 137–152.

Weltzien, F.A., C. Pasqualini, M.E. Sebert, B. Vidal, N. Le Belle, O. Kah, P. Vernier, and S. Dufour. 2006. Androgen-Dependent stimulation of brain dopaminergic systems in the female European eel (*Anguilla anguilla*). Endocrinology 147: 2964–2973.

Winter, H.V., H.M. Jansen, and M.C.M. Bruijs. 2006. Assessing the impact of hydropower and fisheries on downstream migrating silver eel, *Anguilla anguilla*, by telemetry in the River Meuse. Ecol. Freshw. Fish 15: 221–228.

Wooton, R.J. 1984. Introduction: strategies and tactics in fish reproduction. *In:* G.W. Potts and R.J. Wooton [eds.]. Fish Reproduction: Strategies and Tactics. Academic Press, London, UK pp. 1–12.

Yamada, Y., H. Zhang, A. Okamura, S. Tanaka, N. Horie, N. Mikawa, T. Utoh, and H.P. Oka. 2001. Morphological and histological changes in the swim bladder during maturation of the Japanese eel. J. Fish Biol. 58: 804–814.

Yokouchi, K., R. Sudo, K. Kaifu, J. Aoyama, and K. Tsukamoto. 2009. Biological characteristics of silver-phase eels, *Anguilla japonica*, collected from Hamana Lake, Japan. Coast. Mar. Sci. 33: 54–63.

Young, G., B.T. Björnsson, P. Prunet, R.J. Lin, and H.A. Bern. 1989. Smoltification and seawater adaptation in coho salmon (*Oncorhynchus kisutch*)—plasma prolactin, growth hormone, thyroid hormones, and cortisol. Gen. Comp. Endocrinol. 74: 335–345.

CHAPTER 4

Marine Migratory Behavior of the European Silver Eel

Håkan Westerberg

1. The Eel Question

The European eel (*Anguilla anguilla* L.) is an enigmatic and fascinating species. The slippery nature of the eel has been the source of proverbs all over its distribution range, from North Africa to Scandinavia. In fact, the ancient Egyptians used a pictogram of a man holding an eel by the tail to illustrate an impossible task. Much about the life of the eel is also difficult to grasp. From ancient times the reproduction of the eel was the primary Eel Question. Nobody had seen a ripe eel or breeding eels. Numerous theories were put forth, from Pliny's hypothesis that new eel arise from the slime scraped from the bodies of adult eels on stones to the poetic notion of Franciscus Mercurius van Helmont, who believed that eels came from dew falling in the month of May on the banks of ponds and rivers (Walton 1653).

Spontaneous generation in the form of earth worms appearing from "*the earth's guts*" was the explanation given by Aristotle in Historia Animalum. "*Such earthworms are found both in the sea and in rivers, especially where there is decayed matter: in the sea in places where sea-weed abounds, and in rivers and marshes near to the edge; for it is near to the water's edge that sun-heat has its chief power and produces putrefaction. So much for the generation of the eel.*" The authority of Aristotle was strong, but doubts about the true nature of the reproduction of the eel remained. Carl von Linné classified the eel as

Swedish University of Agricultural Sciences, Institute of Freshwater Research, 178 93 Drottningholm, Sweden.
Email: hakan.westerberg@slu.se

viviparous, a view shared by other 18th century naturalists as Leuwenhoeck and Buffon. The empirical foundation for this was weak and may have been due to confusion with the eelpout (*Zoarces viviparus* L.).

The first indication that eels spawned like other fish came in the early 18th century, when Vallisneri, at the University of Bologna, identified true ova in eels from the Italian eel fishery in Comacchio, an observation later verified by Mondini (Mondini 1783). The finding however, was questioned and it took 50 years before the existence of the female reproductive organ in eels was finally recognized. About a hundred years later, the male reproductive organ was described by Syrski (Syrski 1874). Inspired by this, Sigmund Freud searched in vain for sperm in the organ described by Syrski and was so disappointed that he left ichthyology for psychoanalysis (Freud 1877).

Once the reproductive organs were identified, the Eel Question changed from how to where. The autumn migration towards the sea was well known, but the location of the breeding place, or places, was unknown. Grassi had identified the large, transparent leptocephalus larva as the early form of eel larvae (Grassi 1896). Later Johs Schmidt, after accidentally catching the first leptocephalus larva outside the Mediterranean Sea west of the Faroes in 1904 (Petersen 1905), embarked on a lengthy, wide ranging search for ever smaller leptocephali. This was conducted by undertaking several Danish eel expeditions in the Mediterranean and Atlantic, interrupted between 1915 and 1918 due to World War I. The work was resumed in 1920 with two large expeditions, Dana I and II, covering the whole North Atlantic Ocean. In 1922, Schmidt could announce that he had found the breeding place of the eel in the Sargasso Sea (Schmidt 1922).

That the breeding took place 5000 km–8000 km from the European coast came as a surprise. The Eel Question was now about how the eels managed the transatlantic migration. That this was possible was questioned by Tucker, who argued that Schmidt was wrong about the European and American eels being different species, and asserted that the European population was maintained by larvae of American parentage (Tucker 1959). This hypothesis has been disproved beyond doubt by modern genetic analysis, which shows that the European and American eel are different species and that the European population is panmictic, indicating a single breeding place.

But there are still many questions about eel migration that remain unanswered. The present review is an attempt to describe how the study of the marine phase of the spawning migration has been approached specifically for the European eel and what the present understanding is, using the observation techniques as a point of departure and with a special emphasis on the recent results from the EELIAD project. A review by Righton et al. discusses the migration problem from a broader perspective, including physiology and the whole life cycle of all 19 species of *Anguillidae*

(Righton et al. 2012). The standard introduction to all aspects of eel is the monograph by Tesch (Tesch 2003).

2. Early Attempts—Conventional Tagging

The migrating eel stops feeding and cannot in general be captured on hook and line. Its slender body allows it to escape conventional trawls and nets. As a consequence, once the eel leaves the coast it is essentially never caught and conventional tagging is unsuccessful in collecting information about its oceanic migration. As an example, of the thousands of eels tagged in freshwaters in the North Sea region, just two have been recaptured in the sea, both on the northwest coast of Denmark where a pound net fishery exists with the ability to catch eels (Lühmann and Mann 1958).

An exception to this situation is found in the enclosed Baltic Sea. For centuries, a coastal eel fishery using large pound nets, or fish traps called "hommor", has been common in the Baltic. This fishery primarily targets the autumn migration of silver eel both on the eastern and western side of the Baltic proper. In the 19th century, the fishery biologist Rudolph Lundberg was able to get quite an accurate picture of eel migration by observations of how the leader nets of the gear were orientated and how the catch varied in time along the coast (Lundberg 1881). Basically, the migration direction was south and southwest to reach higher salinity water. The season of the fishery also indicated a progress towards the southern Baltic, staring later in the autumn the closer you got to the Baltic outlet.

The next step in investigation of the eel migration in the Baltic was tagging experiments, starting in 1903 in Finland. The purpose of the investigation was to track the route the eels took from Lake Ladoga and to explore the potential for a Finnish coastal eel fishery. A total of 40 eels were released at three places on the Finnish south coast. They were tagged by tying a silken cord, twisted from red and yellow threads, at the dorsal fin (Nordqvist 1903). In the first year, a single eel was recaptured in the Stockholm archipelago, having crossed 350 km to the western side of the Baltic.

At the same time Filip Trybom started tagging eels at several places both on the Swedish east coast, in Finland, and in Germany during four migration seasons (Trybom 1905; Trybom1908; Trybom and Schneider 1908). He used silver plates inserted under the skin (Fig. 1a). The tags were numbered, which made it possible to identify recaptured eels individually. The results verified Lundberg's hypothesis and showed that eels indeed left the Baltic. The main migration route was from the eastern Baltic coast

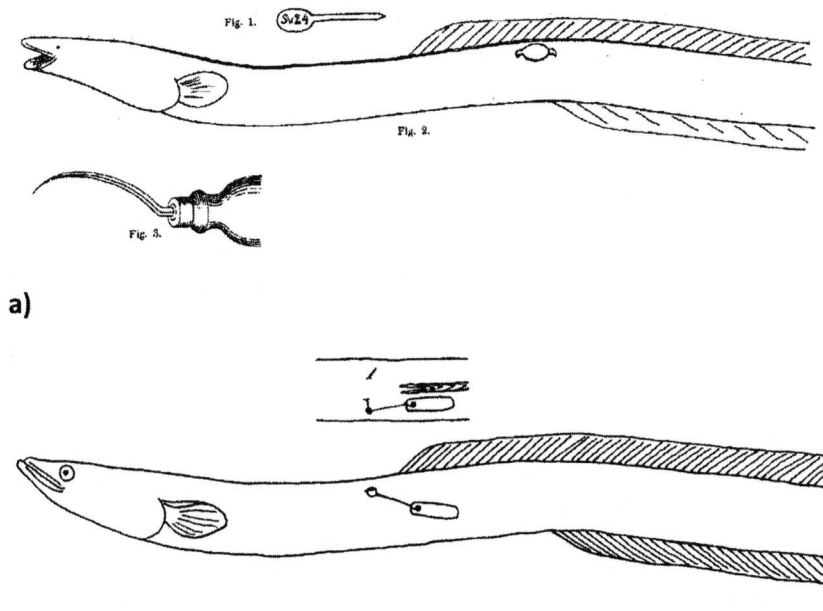

a)

b)

Fig. 1. a) The first tags used on eel. Fig. 1) The engraved silver plate. Fig. 2) Position of the tag on eel. Fig. 3) Instrument used for inserting the tag. From Trybom 1905. b) The Carline tag, used since the 1960s in most tagging studies in the Baltic. The tag is attached with two stainless steel wires under the skin (Illustration Frida Sjöberg).

across the open Baltic and then south along the Swedish side. The migration speed varied, with a characteristic maximum of 20–30 km/day.

Since then, many eel tagging experiments, with a total of > 40,000 eels tagged, have been done in different parts of the Baltic. The majority were tagged using Carlin tags (Fig. 1b). The results have been synthesized, e.g., by Määr, Ask and Erichsen and Svärdsson, all resulting in a more or less similar map of the migration (Fig. 2) (Määr 1947; Ask and Erichsen 1976; Svärdson 1976). A recent meta-analysis was made by Sjöberg and Petersson which confirms the early results (Sjöberg and Petersson 2005).

Some tagging experiments aimed to study the orientation behavior of the eel. Karlsson studied the effect of attaching permanent magnets (Karlsson 1985) and Karlsson et al. compared the migration success of intact and olfactory blocked Polish eels released in the south eastern Baltic (Karlsson et al. 1988). No significant effects were demonstrated, but the low recapture rate observed for the Polish eels, taken from the Masurian lakes where they probably had been stocked as elver from western Europe, became a starting point for Lars Westin to conduct a number of tagging studies to look at the effect of stocking on migration success (Westin 1990; Westin 1998; Westin 2003). His conclusion was that a necessary condition

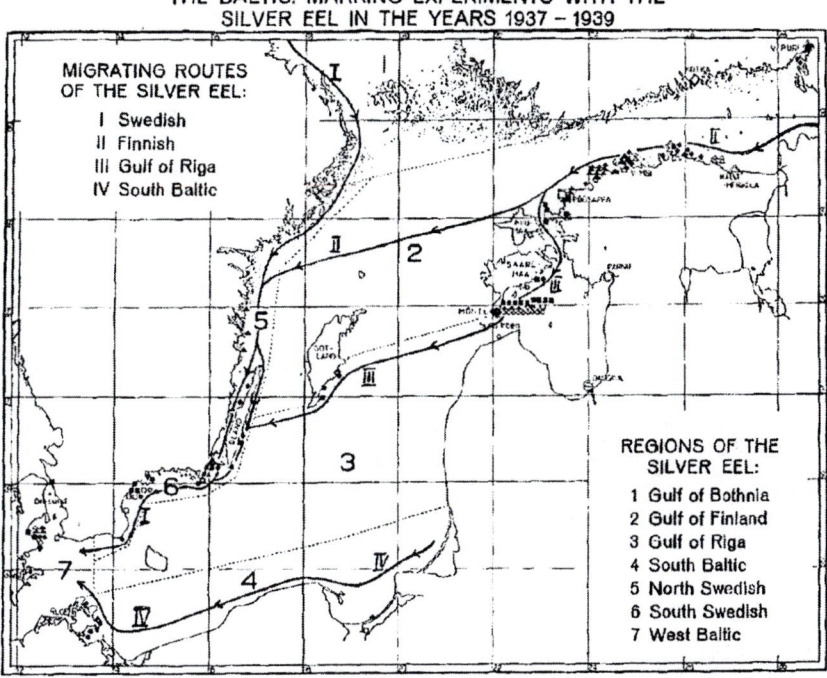

Fig. 2. A generic map of the silver eel migration in the Baltic, based on taggings made in the early 20th century. From Maar 1949.

for a successful migration was the olfactory orientation cues that had been imprinted during the Atlantic journey of the juvenile eel.

A limitation of the conventional tagging technique is that only the start and endpoints of the eel's trajectory are revealed. Everything in between, as horizontal meandering or swimming depth, is unknown. To solve this problem, a modified tagging method was tried by Edelstam in his investigation of a possible star-compass mechanism in the eel (Edelstam 1965). He used small balloons attached to the eel with thin treads. With the help of a light on the balloon he could follow the eels at night from a boat and note their vanishing direction. My own experience with this technique is that it is only a matter of time before the balloon is towing the eel downwind rather than the eel towing the balloon. Detailed study of eel migration behavior had to wait for electronic tagging methods.

3. Telemetry Tracking

Telemetry technique using ultrasonic pingers attached to a fish to locate, follow, and optionally transmit environmental or physiological data, started in the mid-1950s (Stasko and Pincock 1977). Initially, the tags were bulky and only suitable for large fish as adult salmon or sharks. The technology improved and tags became gradually smaller. The first trials with acoustic telemetry on silver eel were made in 1969 in the southern North Sea by the renowned eel expert Friedrich-Wilhelm Tesch and showed a northward swimming direction relative to the water (Tesch 1972). The direction over ground was deflected east or west by the tidal currents. A follow up experiment with silver eels in the same area showed the same northerly direction (Tesch 1974a). In this study, yellow eels were tracked as well, showing homing in a south easterly direction relative to the water. To pursue the investigation of homing and possible geomagnetic navigation, transplanted yellow eels were tracked in the same area (Tesch 1975). This showed the same south-east direction, different from the expected northeast direction to the home area of the eels.

All the early telemetry tracking studies were made on shelf seas, albeit in open sea conditions, and often lasting less than one day (Tesch 1979; Westerberg 1979; Tesch et al. 1991; McCleave and Arnold 1999). Some longer tracks in the Baltic (Westerberg 1979) showed a diurnal behavior, with the eel resting at the bottom in daytime (Fig. 3). A study in the North Sea using sector-scanning sonar and transponders on the eel was made by McCleave and Arnold (McCleave and Arnold 1999). In two out of thirteen cases, the behavior was interpreted as a selective tidal stream transport, i.e., silver eels taking advantage of advection by the tidal current by alternating between swimming in mid-water on one tidal phase and resting on or close to the bottom on the opposed tide. There were other cases with a clear diurnal vertical migration similar to what was seen in the Baltic, with the eel active in mid-water and near the surface during the night and deeper and at the bottom in the day. In some of the eels however, the diurnal pattern was reversed. The first ocean tracking studies were made by Tesch (Tesch 1978). The main finding from those still very short tracks was that the eels moved in the upper 100 m during the night but dived to depths below 400 m, which was the limit of the pressure sensor, at sunrise. A similar diurnal behavior was later seen in the Mediterranean Sea using tags with larger pressure range (Tesch 1989). The maximum daytime depth was 600–700 m, which was close to the bottom depth in the study area.

The short duration of the tracks and the varying and conflicting results makes it difficult to generalize about migration behavior. No universal direction preference is evident. In all cases with pressure sensing tags, frequent diving activity was noted, but the short tracks don't allow a clear picture of the diurnal pattern. Taken together, the meager results from active telemetry tracking of eels at sea and the high cost and effort to conduct such

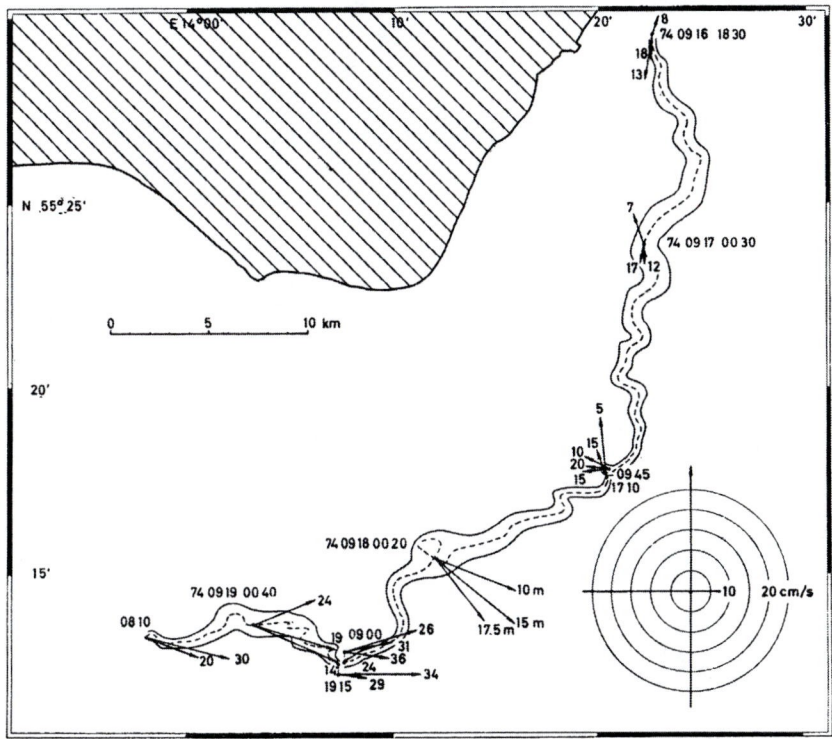

Fig. 3. Plot from an early telemetry experiment with silver eels in the Baltic. The track covers three days. The envelope around the dotted eel trajectory shows the 63% uncertainty limits of the Decca fixes. Arrows show current measurements with the depth indicated at the tip and time of measurement at the origin. Note the periods where the eel rests at the bottom during daytime. From Westerberg 1979.

studies led to an effective cessation in the 1980s. Passive tracking using individually coded acoustic tags and fixed arrays of hydrophone buoys is now the preferred alternative for coastal studies (Aarestrup et al. 2008; Aarestrup et al. 2010; Davidsen et al. 2011; Westerberg et al. 2008).

A way to bypass those difficulties and get data from the spawning area was to transport hormone treated eels to the Sargasso Sea and release them there. This was tried during a reconnaissance cruise to the Sargasso in 1979 (Tesch et al. 1979). Four eels were tracked from one to 12 hours. The directions were random and the maximum depth reached by the eels was approximately 700 m. A similar experimental design was tried by Fricke and Kaese (Fricke and Kaese 1995). In this case, two eels were followed for four and seven hours in daytime. The swimming depth was 200–300 m, which was interpreted as a preference for the boundary between the surface water mass and the North Atlantic central water.

4. New Tracking Technology

If tracking at sea was impractical, an alternative was to simulate the ocean migration in the laboratory. This approach has been developed successfully by a group of researchers in the Netherlands (van Ginneken et al. 2000; van Ginneken et al. 2005; van den Thillart 2004). They used swim tunnels where eels were studied during a forced swimming over 5,500 km. Those studies disposed finally, the doubts put forward by Tucker in 1959 that the energy reserve of the European eel was insufficient for a transatlantic migration (Tucker 1959). The swim tunnel experiment included measurements of oxygen consumption and the change in body composition. The eels showed a remarkably high swimming efficiency compared to other fish.

Several other aspects of the physiology of eel migration have been studied by this technique, e.g., the effect of swimming on maturation (van Ginneken et al. 2007) and the negative effect of infection with the swim-bladder parasite *Anguillicola crassus* on swim capacity and endurance (Palstra et al. 2007).

The high cost and low information return of active tracking in the ocean lead this technique to a dead end. Not until new technology with electronic data storage tags (DST) and satellite pop-up tags (PSAT) in small sizes became available was the study of eel migration behavior in the ocean resumed. DST tags are available in different sizes and with a range of sensors to monitor internal or external environment of the fish. A modern DST can store in excess of a million data points and weigh 1 g–2 g in water (Metcalfe et al. 2009). The limitation is that the tag must be recovered for the data to be downloaded. PSATs on the other hand are larger and relatively expensive, but have the advantage that a high proportion will deliver data via a satellite link when they detach from the fish and surface (Block et al. 1998). The amount of data that can be transmitted is limited and has to be pre-processed in the tag. Geolocation is made by the processor in the tag by measuring daylight at a high sampling rate to monitor the time of sunrise and sunset. With this information the time of local noon and the day-length can be estimated, which is used to calculate longitude and latitude respectively. This requires a certain amount of light and hence that the animal moves relatively close to the water surface. Unfortunately, the diel vertical diving behavior and great swimming depth means that this method does not work for migrating eels.

DST was first used in the Baltic, where 16 silver eels were tagged on the Swedish east coast. Half of those were recaptured after an average of 14 days at large (Westerberg et al. 2007). The detailed swimming depth time series show a consistent diurnal pattern with the eel resting at the bottom from shortly after sunrise to just before sundown (Fig. 4). Using the daytime bottom depth and daily mean swimming velocity, the trajectories

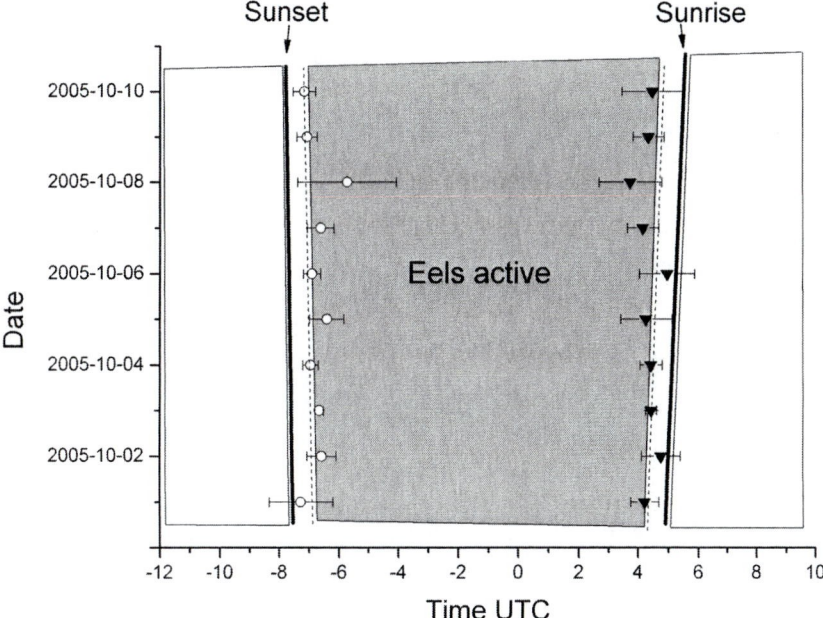

Fig. 4. Average time for onset (open circle) and termination (filled triangle) of diurnal activity for six silver eels tracked with DST in the southern Baltic. The error bars show the 95% confidence interval and the dashed lines the time of start and end of civic twilight at the actual date and longitude. Redrawn from Westerberg et al. 2007.

were reconstructed, showing a migration following the coastline over 130 km–170 km.

Satellite tagging was used at the Danish Galathea 3 expedition in 2006 (Aarestrup et al. 2009). PSAT had been used earlier on the large, long-finned eels (*Anguilla dieffenbachii*) in the Pacific Ocean (Jellyman and Tsukamoto 2002) but this was the first time they were tried with the smaller European eel. Twenty-two eels were released on the west coast of Ireland and transmission was received from 14 of those. All but one track were terminated before the programmed date, which means that the tags were detached from the eel, either due to predation or to a failing attachment. The general direction was southwest for most tags and the longest distance between release and first satellite location was 1,340 km.

The most striking observation was the very regular diel vertical migration once the eels had left the continental shelf. Figure 5 shows a short episode recorded over the same time period for two of the eels. Those two eels were most probably migrating independently as they left the continental shelf with a separation of four days and had been migrating for approximately 30 days in the mesopelagial before the period shown in

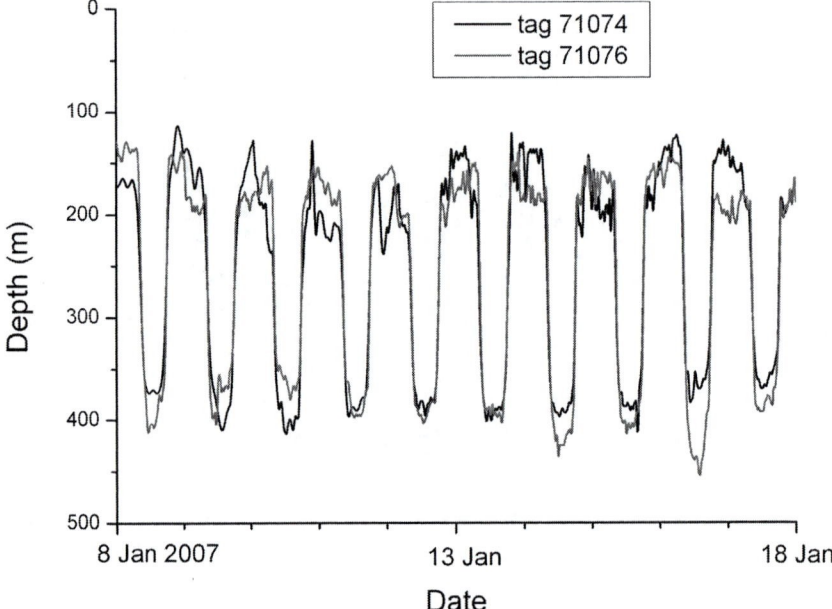

Fig. 5. The simultaneous recordings of swimming depth from two PSAT tagged eels in the area off the Irish shelf. Partly redrawn from Aarestrup et al. 2009.

the diagram. The swimming depth changes are nevertheless essentially synchronous.

The observations by Westerberg et al. demonstrated a strong relation between the day-night cycle and the time of sunrise and sunset (Fig. 4) (Westerberg et al. 2007). It is a plausible assumption that the same applies to the diel vertical migration seen in the open ocean, which means that the daily depth cycle can be used to calculate the time of local noon and thus the longitude of the eel along the track, even without actual daylight measurements. This geopositioning method has become an important tool in the later analysis of DST and PSAT data.

5. EELIAD

An ambitious EU funded eel research project—EELIAD (European Eels in the Atlantic: Assessment of their Decline)—started in 2008. The project engages researchers and managers from seven European countries and 12 institutions. The main objectives are investigation of the marine migrations of European eels in relation to regional ecological features, and assessment of the most important factors that influence silver eel production and migration success.

A major part of the EELIAD project is a tagging program where a total of 161 eels have been tagged using PSAT (Microwave Telemetry, X-tag) and 512 eels with DST (CEFAS Technology Ltd., G5 tags). Releases have been made in France, Ireland, Spain, and Sweden to cover different parts of the eel distribution range. The DST were made buoyant and recovery relies on the tags drifting ashore and being found there. This method shows a surprisingly high recovery rate of the floating tags, even higher than what is usual for conventional tagging in commercial fisheries.

The project is not yet terminated and DST tag data are still coming in. As of August 2012, data from a total of 80% of the PSAT have been received by the ARGOS system. The DSTs are of two kinds: mounted externally with a release mechanism or implanted in the body cavity. The return of external tags is 19% and internal tags, 9%. In total, 194 datasets have been obtained with 6,900 days of eel behavior from the open ocean and an additional 4,700 days from Lake Mälaren and the Baltic Sea.

Analysis of this wealth of data is far from complete and so far, little has been published. Preliminary results were presented at the World Fisheries Congress 2012 and some highlights will be given here. The major findings so far are:

- The ubiquitous, large daily vertical migration of eels, uninterrupted over extended periods.
- The common route taken by eels tracked from the Baltic-Kattegat area north of the Shetland or Faroe Islands before turning south.
- The high incidence of predation on the migrating eels.
- The identical migration behavior of eels that have been recruited naturally and eels that were translocated as glass eels.

One eel exemplifies many of those points and was also one of the first recoveries, named the "Shetland tag" by the EELIAD project. This eel was tagged in November 2008 in the Sound between Sweden and Denmark. The tag, an implanted DST, was found on North Beach in Shetland on March 2009. The maximum swimming depth and the longitude estimated from the diurnal depth changes restricts the possible positions along the track. This shows that the eel must have migrated along the Norwegian Trench, which is a deep channel from Skagerrak running north along the Norwegian coast and opening into the deep basin of the Norwegian Sea at about 62°N.

Judging from the temperatures encountered at depth in the Norwegian Sea, the eel first seems to have continued north to at least 63°N, where the temperature was 1.2°C at 400 m. It then turned south and crossed the shallow ridge between Scotland and the Faroe Bank on 12th February, 2009. This is seen from the sudden change in deepwater temperature—from approximately 2°C at 500 m in the Shetland channel to 8°C at 600 m in the Atlantic water south of the ridge (Fig. 6).

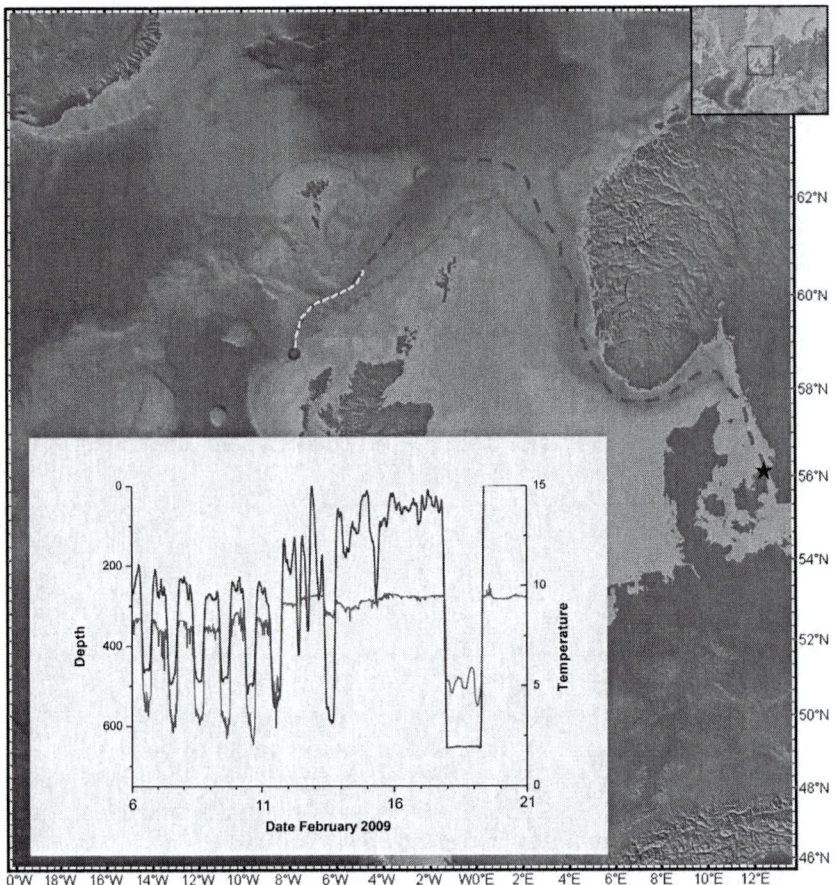

Fig. 6. The reconstructed trajectory of a silver eel tagged internally with a floating DST at the outlet of the Baltic (star). The position where the tag surfaced is shown as a filled circle. The inserted swimming depth and temperature diagram shows the portion of the track highlighted in white. For discussion see text.

Color image of this figure appears in the color plate section at the end of the book.

On 14th February, the diurnal cycle disappeared and the depth became shallower and highly irregular. On 17th February, the tag dropped to the bottom at 640 m depth and remained there for 34 hours before it rose to the surface and started drifting. The position where this happened could be determined fairly accurately by combining the bathymetry of the area and the sea surface temperature measured by satellites on this date. The result was 59°N and 7.6°W, giving a total trajectory of more than 1,900 km.

The behavior shift on 14th February could either be a sign that the eel was exhausted and died, or that the eel had been predated, swallowed, and voided four days later. The same tag was definitely predated somewhat later while it was drifting at the surface (Fig. 7). The temperature had suddenly risen to approximately 39°C. The fact that the pressure remained unchanged indicates that the tag was swallowed, most probably by a bird, and possibly flown to the Shetland Islands.

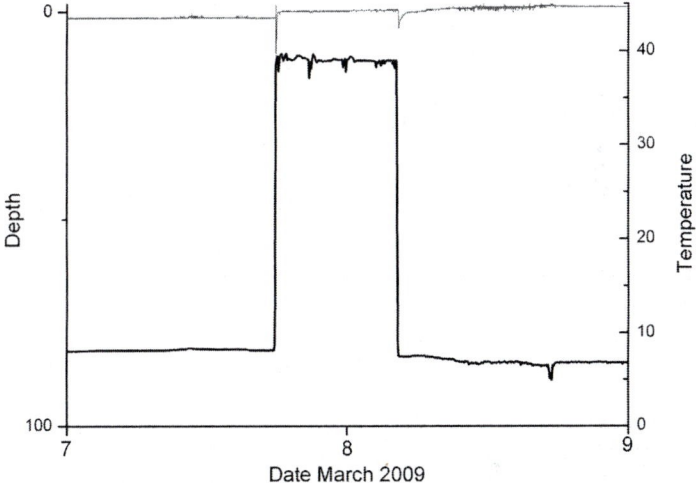

Fig. 7. Temperature (black) and depth (grey) recordings obtained from the same tag as in Fig. 6, showing the tag likely being swallowed by a bird while drifting.

During the period 2008–2010 a total of 170 silver eels have been tagged on the Swedish west coast. The eels were either captured in two rivers, one discharging in Kattegat and one in Skagerrak, or in the Sound on leaving the Baltic. Fifteen eels were fitted with PSAT, the rest in approximately equal number with internal and external DST. So far 19 data records with more than one month of data have been obtained, giving a total of 1,950 days of detailed eel behavior.

The trajectories of those eels have been reconstructed using a combination of different methods:

- Longitude estimates from the daily diving cycle. The local noon is found as the mid-point between the times of maximum descent and ascent and this is used to calculate the longitude, correcting for the equation of time for the current date.
- The maximum recorded swimming depth on the current date. The bathymetry of the general area where the eel could have been restricts the possible position.

- The observed temperature at different depths, which restricts the possible position further using comparison with oceanographic mean climatology, e.g., the World Ocean Atlas or the WOCE database (Gouretski and Koltermann 2004).

In addition, some specific features seen in most records were used as fix points. One was mentioned above, the point when there is a sudden change in environmental temperature from less than 2°C to more than 7°C during the deep part of the diurnal cycle. This corresponds to when the eel passes the Island-Scotland ridge. The popup position could also be determined with relatively high accuracy using the longitude estimate from the end of the track and the measured sea surface temperature where the tag popped up, compared to satellite SST charts for the current date.

The main result is that all 19 eels behaved in essentially the same way as described for the "Shetland tag", following the Norwegian Trench, first west and then north along the Norwegian coast. In the Norwegian Sea the trajectories are more varying with some meandering, but subsequently the eels turn south and southwest. Most eels pass over the Wyville-Thomson ridge in the Faroe-Scotland passage but one in four crossed the Iceland-Faeroe rise west of the Faeroe Islands. In the Atlantic water south of the Iceland-Scotland ridge the eels move in a south-westerly direction in deep water outside the continental shelf.

The amplitude of the daily swimming depth cycle varies both between individuals and at different places along the trajectory. Evidently the maximum depth is restricted by topography in the Norwegian Trench, where typically the eels move between 100 and 300 m, occasionally with periods at the bottom. In the Norwegian Sea, the maximum swimming depth increases to >500 m but remains far above the 2,000 m bottom depth. The same mesopelagic behavior is seen west of Scotland and Ireland.

The deep water of the Norwegian Sea basin is cold. In this area all eels were exposed for long periods to temperatures less than 2°C during the daytime, deep swimming phase. In some cases the eels even migrated at slightly sub-zero temperature (Fig. 8). This differs from what has been found in laboratory studies of activity (Westin and Nyman 1979), where silver eel activity ceased completely at 1–2°C.

The average migration speed has been calculated from the beginning of active migration to the popup position. The reconstructions of the trajectories are not accurate enough to resolve changes in swimming speed along the track in any detail, especially in the Norwegian Sea where the latitude estimates are most uncertain. The calculated overall migration speed becomes a lower estimate as small scale meanders have been smoothed. The average length of the individual trajectories was 1,700 km (the longest was 3,000 km) and the mean migration speed was 22.4 km/day or approximately 0.3 BL/s.

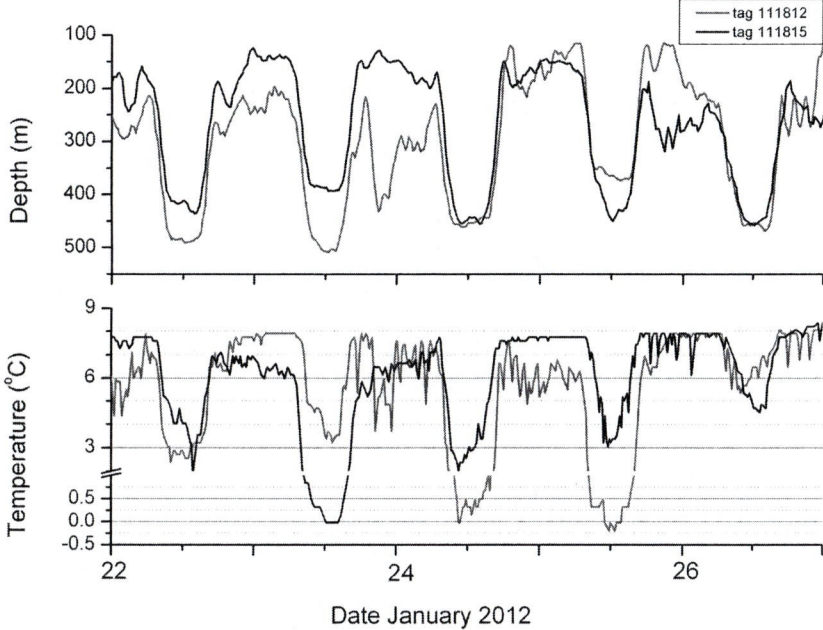

Fig. 8. Simultaneous recordings of swimming depth and temperature from two PSAT tagged eels migrating in the Norwegian Sea north of the Faroe Islands. Note the below zero temperatures encountered while the eels move at extreme depths in daytime.

A controversial issue in the current management of the European eel is the practice of stocking. This is a major measure in several national management plans. The effect of stocking on recruitment has been questioned however, mainly based on the results of the tagging experiments in the Baltic (Westin 1998; Westin 2003; Svedäng and Gippert 2012). The translocation of eels is seen as problematic, as translocated (i.e., stocked) eels, according to these studies, display impaired navigational abilities.

To test this hypothesis, tagging in 2010 and 2011 was done on eels from two river systems where the recruitment history of the eels were known, one where no known stockings had taken place and one above some 10 hydropower dams obstructing immigration and where massive stocking with imported glass eel has been practiced since 1990. An equal number of eels were tagged from the two rivers and released together at the coast.

The majority of the long data records discussed above comes from those releases. The results cannot verify any migration behavior difference between naturally recruited and translocated eels. Migration speed and the route along the Norwegian trench is the same. Indeed, the simultaneous diurnal depth changes are more or less identical for the two eels shown in Fig. 8. Both eels where migrating somewhere in the Norwegian Sea, approximately two months after release, one of stocked origin and one recruited naturally.

The releases of tagged eels in Ireland, France, and Spain have not produced the same coherent picture of the oceanic migration as what was found for the northern migration route. The main reason is the very poor data return. Essentially all external tags were prematurely detached and the median length of the records of eel behavior data was approximately seven days. This is too short to determine a migration route.

The reason for the rapid loss of tags is not fully known. Loss of the tag due to rejection or mechanical failure of the attachment cannot be excluded for the externally attached tags. A 6-month laboratory study of the attachment methods used in the project, however, showed a high survival and retention rate (Økland et al. 2013). A more likely explanation is exhaustion or unexpected high predation pressure on the tagged eels.

In more than 20 percent of the PSAT recordings, predation could be confirmed. The Microwave Telemetry X-tag stores the highest and lowest measurements of the daylight level for each day. If the maximum light level was the same as the minimum, even if the tag showed movements in the upper layer, it can be concluded that the tag had been swallowed. In a few cases, even the identity of the predator could be determined. Marine mammals were involved in at least four cases. This was concluded from the body temperature recorded while the tag was in the stomach. Another predator was a warm-bodied shark, which in this area only could have been a porbeagle or a thresher shark (Fig. 9). A high incidence of eel predation by porbeagle sharks have been seen for the American eel in the Gulf of St. Laurence (Béguer et al. 2012).

6. Hypotheses about Orientation and Navigation

Animal navigation in general and fish migration in particular is a fascinating subject where much still is unknown. The very long spawning migration of the European eel has generated several hypotheses about how eels find their way.

Basically animal navigation can be classified into:

- Orientation, where the animal either follows some specific feature of the environment, e.g., an ocean front or the shoreline, or possesses a compass mechanism and tries to follow a specific heading, which can be innate or determined from local environmental clues.
- Navigation, where the animal, in addition to a compass mechanism, has a map to tell it where it is and which direction to take to reach the goal.

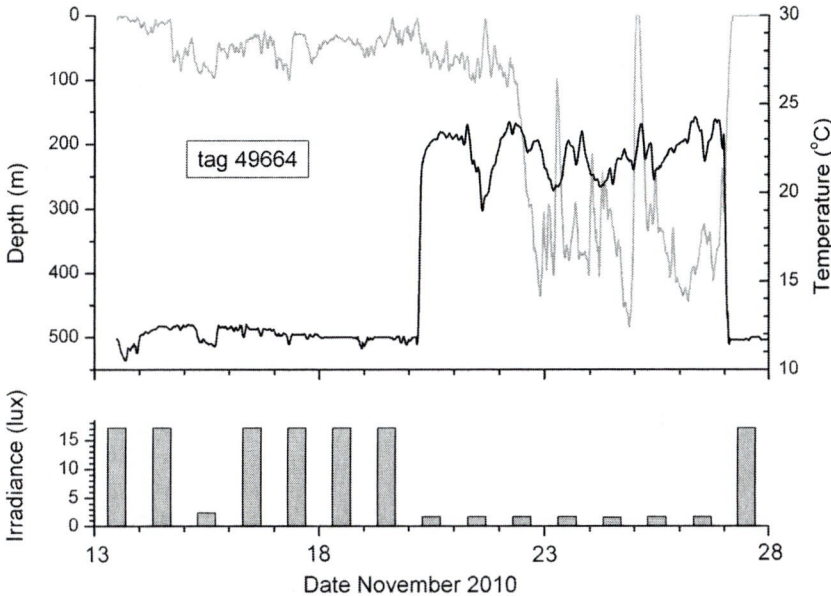

Fig. 9. Temperature (black) and depth (grey) recorded from a PSAT tagged eel released on the Irish west-coast and predated by a warm-bodied shark (upper panel). The lower panel shows the maximum daylight irradiance recorded for each day. The output 2 lux corresponds to dark conditions; 17 lux is the saturation value of the sensor.

Attempts to explain how the European eel manages the transatlantic migration have involved both orientation and navigation. The early tagging results in the Baltic were interpreted as a simple compass mechanism with an innate southwest direction, which was deflected when the eel encountered a coastline (Määr 1949). A similar southwest preference was proposed by van Ginneken et al. (van Ginneken et al. 2005a). However, such a simple vector orientation is difficult to apply for the whole distribution range of the eel, which requires different innate directions from different starting points. The European eel is panmictic and there are no genetic differences between eels across the whole range. There has been a controversy about whether this classic view of panmictia holds, but recent results (Als et al. 2011) have resolved the issue. Dependence on a compass sense thus requires a map and true navigation or some modifying cues to switch between different directions depending on where the eel is located.

F-.W. Tesch and his co-workers have in a series of laboratory experiments studied the directional choice of eels in natural and artificial magnetic fields and how this is modified by life stage, salinity and other environmental factors (Tesch and Lelek 1973; Tesch 1974b; Karlsson 1985; Tesch et al. 1992). The results are conflicting and not always easy to interpret. The main hypothesis is that eels possess a geomagnetic compass sense and that the preferred direction changes with salinity and possibly other environmental factors, which allows a sequence of directions taking the eel to the goal.

Sensitivity to the geomagnetic field has also been claimed by Durif (Durif and Skiftesvik 2007; Durif pers. com.). She reports sensitivity both to direction, inclination, and intensity of the field. A hypothesis put forward is that silver eels, once they are out in the Atlantic, can orient towards lower magnetic field areas, i.e., towards the equator, until the right field strength has been found, and then follow an isoline towards the Sargasso Sea. This can be classified as a true navigation mechanism with a magnetic map (Lohman et al. 2007).

Magnetosensitivity has also been found in the Japanese eel (*A. japonica*) (Nishi et al. 2004). Experiments made with the American eel (*A. rostrata*) were, on the other hand, negative (Rommel and McCleave 1973a). Rommel and McCleave instead demonstrated that *A. rostrata* had a very high sensitivity to weak electric fields and proposed another orientation mechanism, that eels could detect the electric fields induced by ocean currents flowing in the geomagnetic field and in this way be able to follow the path of the Gulf stream to the Sargasso Sea (Rommel and McCleave 1973b). However, it has not been possible to reproduce the same sensitivity to electric fields in experiments with European eels (Enger et al. 1976).

The alternative to the mechanisms depending on the geomagnetic field discussed above is orientation to some other clue. Based on laboratory activity studies and a few telemetry tracks, Westin and Nyman suggested that a single stimulus, temperature, was enough to explain the orientation (Westin and Nyman 1977; Westin and Nyman 1979). Colder temperature causes an avoidance reaction and increased activity. Combined with a horizontal temperature gradient, this could guide the eel in an appropriate direction. Westin later revised this hypothesis to include olfaction as an important clue (Westin 1990; Westin 1998). He found great differences in migration success between eels made anosmic by blocking the nostrils and untreated natural eels. Telemetry tracking of anosmic eels (Tesch et al. 1991) also showed irregular swimming without a common direction and approximately half the swimming speed of normal eels in the same study. This led Westin to the hypothesis that there is a sequential imprinting on odors during the immigration as larvae and juveniles and that the spawning migration uses this sequence for orientation back.

A fundamental problem with migration mechanisms based on following hydrographic features, be they temperature, induced electric fields or odor, is that horizontal gradients are on a very large scale and it is difficult to see how an animal can detect minute differences in, say, temperature over tens or hundreds of kilometers. Turbulent eddies will also tend to distort the horizontal gradient locally. There is also a lack of visual points of reference making rheotactic orientation impossible (Arnold 1981). A way to avoid this difficulty is to focus on the vertical distribution of properties instead of the horizontal (Westerberg 1984).

The basic concept is that information can be found locally in the vertical distribution of a property rather than in the complex, large-scale distribution. All naturally stratified water bodies have strongly anisotropic structure where the vertical scale mirrors the horizontal water-mass distribution on a much compressed scale (Iselin 1939; Wunsch and Ferrari 2004). This means that a vertical search, like the sporadic short period dives with amplitude of the order of magnitude 100 m commonly seen in the eel, could be a search for a layer containing an odorant or other property specific to the spawning area. The vertical zigzag movement could be an analogue to the horizontal zigzag of salmon following the interface between water from different tributaries in a spawning river (Johnsen and Hasler 1980). The directional clue used by the eel is unknown. A likely clue is the local direction of the current shear (Westerberg 1984).

Harden Jones in his book on fish migration wrote, "...*one of the fundamental questions to be asked, and answered, is simply this: what are the movements of migrants relative to those of the water at the depth at which they are swimming?*" (Harden Jones 1968). The data gathered by the new remote tracking technology is approaching this goal, but there is still a long way to go before we can finally decide between the possible orientation mechanisms. In light of the EELIAD results, a simple innate direction seems highly unlikely and a sequential imprinting of the route from the Sargasso to the continental growth area also seems to be ruled out. The nature of the migration—a true navigation using map and compass or orientation along some oceanographic feature—remains an open question, as well as the roles of learning and innate responses.

7. Future Prospects

What is the Eel Question of today and will tracking technology be able to give an answer? There are now many eel questions, e.g., the mechanism of migration discussed above. The most pressing however, are the management questions in the light of the recruitment crisis of *Anguilla* worldwide (Casselman and Cairns 2009). Electronic data storage and satellite tags have gathered much new data; the possibilities with those tracking methods are certainly not exhausted and there is still a lot of analysis to be done on existing data.

Regarding the management aspects, one question is whether eels from the whole distribution range contribute equally to the recruitment. This is important in allocating management measures where they have maximum effect. As was highlighted in a study by Kettle et al., it may well be that the recruitment depends critically on the success of the eels having the shortest distance to the spawning area (Kettle et al. 2011). This type of question was at the core of the EELIAD project, where the basic experimental design was to compare the migration success from different parts of the distribution range. In this respect the project has failed, even if it has yielded many other valuable new insights into the "big blue box".

The main reason for the failure is the unexpected high tag loss due to predation or other causes. The tag loss may at least partly be due to failure of the attachment and this should be studied and improved (Økland et al. 2013), but it may well be that the predation pressure on migrating eels is much higher than what is known. In that case, the only option to study regional differences in migration success is to increase the number of tagged eels substantially. This means higher tagging cost and it becomes of interest to look at the cost-benefit of the different tagging methods.

Just based on the price of the tag, the cost per day of eel data in the EELIAD project (excluding tagging in lakes and the Baltic) was approximately 60 Euro, both for implanted and externally attached DST. The equivalent cost for PSAT was 160 Euro. Other tagging costs will be approximately the same for all tag types, but there is evidently an advantage in using DST tagging amounting to 100 Euro per day of data. By improving the design of the release mechanism and flotation of the externally attached DST, this can become the most effective tagging method to study the relative success of migration from different parts of the range.

To get a deeper understanding of the mechanism of migration will require dedicated experiments to differentiate between the different hypotheses. Tagging methods are important for this as well. The Harden Jones question cited above is still very relevant. The picture we get from reconstructing trajectories is very coarse and the movement of the eel relative to that of the water is still largely unknown. To improve this we should include new sensors in the tags to monitor, say, acceleration and compass direction. Such sensors are available already and it is a question of incorporating them into a floating tag that can be carried by an eel.

Another way of increasing the understanding of migration mechanisms is tagging in combination with sensory deprivation. This necessitates some animal welfare considerations, but this is already part of any tagging study. The experiments that have been done with anosmic eels are interesting but need to be followed up in more detail and in an oceanic environment. There are two main theories about the mechanism of a geomagnetic sense. In one theory, the geomagnetic forces on magnetite particles are the basis (Moore and Riley 2009). The other, more recent model localizes the geomagnetic sense to the eye, where a quantum effect on the pigment cryptochrome allows the animal to "see" the magnetic field (a popular overview is given by Hore) (Hore 2011). The eel may be a suitable subject to study this in fish.

To hold the eel by its tail is a saying used to describe an impossible task, but there is another less known proverb—*Folio ficulno tenes anguillam*—meaning, holding an eel in a fig-leaf. According to Erasmus' "Adages" this originates from the Greek philosopher Diogenianus and Erasmus quotes a poem by Alciato to explain its meaning. *"For a long time wherever you fled, I pursued you: but now you are finally trapped in my snare. No longer will you be able to elude my power. I've caught the eel in a fig-leaf."*

Will future tagging methods give us the fig-leaf to get a grip on the elusive Eel Question?

Acknowledgments

This review is part of a study funded by Grant Agreement GOCE-2008212133 (EELIAD) of the European Union FP7 research program on Environment (including climate change) and prepared under project number 212133. I am grateful to Kim Aarestrup, David Righton, Niklas Sjöberg, and Karin Westerberg for valuable comments and suggestions.

References

Aarestrup, K., E.B. Thorstad, A. Koed, N. Jepsen, J.C. Svendsen, M.I. Pedersen, C. Skov, and F. Økland. 2008. Survival and behavior of European silver eel in late freshwater and early marine phase during spring migration. Fish. Manage. Ecol. 15: 435–440.

Aarestrup, K., F. Økland, M.M. Hansen, D. Righton, P. Gargan, M. Casonguay, L. Bernatchez, P. Howey, H. Sparholt, M.I. Pedersen, and R.S. McKinley. 2009. Oceanic spawning migration of the European eel (*Anguilla anguilla*). Science 325(5948): 1660.

Aarestrup, K., E.B. Thorstad, J.C. Svendsen, N.J. Jepsen, A. Koed, M.I. Pedersen, and F. Økland. 2010. Survival and progression rates of European silver eel in late freshwater and early marine phase. Aqua. Biol. 9: 263–270.

Als, T.D, M.M. Hansen, G.E. Maes, M. Castonguay, L. Riemann, K. Aarestrup, P. Munk, H. Sparholt, R. Hanel, and L. Bernatchez. 2011. All roads lead to home: panmixia of European eel in the Sargasso Sea. Mol. Ecol. 20: 1333–1346.

Arnold, G.P. 1981. Movements of fish in relation to water currents. In: D.J. Aidley [ed]. Animal Migration. Society of Experimental Biology Seminar, Series 13, Cambridge University Press, Cambridge pp. 55–79.

Ask, L. and L. Erichsen. 1976. Blankålsmärkningar vid svenska östersjökusten 1941–1968. Medd. Havsfiskelaboratoriet, Lysekil. Nr. 199.

Béguer, M., J. Benchetrit, M. Castonguay, K. Aarestrup, S.E. Campana, M.J.W. Stokesbury, and J.J. Dodson. 2012. Shark predation on migrating adult American eels (*Anguilla rostrata*) in the Gulf of St. Lawrence. PLoS ONE, in press.

Block, B.A., H. Dewar, C. Farwell, and E.D. Prince. 1998. A new satellite technology for tracking the movements of Atlantic bluefin tuna. Proc. Nat. Acd. Sci. USA 95: 9384–9389.

Casselman, J.M. and D. K. Cairns. 2009. Eels at the edge: science, status, and conservation concerns. Am. Fish. Soc. Symp. 58.

Davidsen, J.G., B. Finstad, F. ØKland, E.B. Thorstad, T.A. Mo, and A.H. Rikardsen. 2011. Early marine migration of European Silver Eel *Anguilla anguilla* in Northern Norway. J. Fish Biol. 78: 1390–1404.

Durif C.M.F. and A.B. Skiftesvik. 2007. Europeisk ål. In: E. Dahl, P. K. Hansen, T. Haung and Ø. Karlsen [eds.]. Kyst og havbruk 2007, pp. 88–89.

Edelstam, C.G. 1965. Long range navigation in animals. Final Report Grant AF EOAR 63-6, Stockholm, OAR-USAF.

Enger, P.S., L. Kristensen, and O. Sand. 1976. The perception of weak electric D.C. currents by the European Eel (*Anguilla anguilla*). Comp. Biochem. Physiol. 54A: 101–103.

Freud, S. 1877. Beobachtungen über Gestalung und feineren Bau der als Hoden beschreibenen Lappenorgane des Aals. Sitzungsb. d. k. Akad. D. Wissensch., Wien LXXV pp. 419–430.

Fricke, H. and R. Kaese. 1995. Tracking of artificially matured eels (*Anguilla anguilla*) in the Sargasso Sea and the problem of the eel's spawning site. Naturwissenschaften 82: 32–36.

Gouretski, V.V. and K.P. Koltermann. 2004. WOCE Global Hydrographic Climatology. A Technical Report. Berichte des Bundesamtes für Seeschifffahrt und Hydrographie Nr. 35/2004.

Grassi, G.B. 1896. The reproduction and metamorphosis of the common eel (*Anguilla vulgaris*). Proc. R. Soc. 60: 260–271.

Harden Jones, F.R. 1968. Fish migration. Edward Arnold Ltd., London.

Hore, P. 2011. The quantum robin. Navigation News Nov/Dec 2011 pp. 17–21.

Iselin C.O'D. 1939. The influence of vertical and lateral turbulence on the characteristics of the waters at mid-depths. Trans. Am. Geo-phys. Union 20: 414–17.

Jellyman, D. and K. Tsukamoto. 2002. First use of archival transmitters to track migrating freshwater eels *Anguilla dieffenbachii* at sea. Mar. Ecol. Prog. Ser. 233: 207–215.

Johnsen, P.B. and A.D. Hasler. 1980. The use of chemical cues in the upstream migration of coho salmon, *Oncorhynchus kisutch* Walbaum. Journal of Fish Biology 17: 67–73.

Karlsson, L. 1985. Behavioral responses of European silver eels (*Anguilla anguilla*) to the geomagnetic field. Helgoland Mar. Res. 39: 71–81.

Karlsson, L., R. Bartel, and F.-W. Tesch. 1988. Migration and orientation of tagged silver eels released in the southeastern Baltic. International Council for the Exploration of the Sea (ICES). C.M. 1988/M17.

Kettle, A.J, L.A. Vøllestad, and J. Wibig. 2011. Where once the eel and the elephant were together: decline of the European eel because of changing hydrology in Southwest Europe and Northwest Africa? Fish Fish. 12: 380–411.

Lohmann, K.J., C.M.F. Lohmann, and N.F. Putman. 2007. Magnetic Maps in Animals: Nature's GPS. Journal of Experimental Biology 210: 3697–3705.

Lundberg, R. 1881. Om ålfisket med s. k. hommor vid svenska östersjökusten samt Öresund. Lantbruksakademins Handlingar och tidskrift, 20: 301–317. Translation available in Report of the commissioner of fish and fisheries for 1883, Washington 1885 pp. 415–430.

Lühmann, M. and H. Mann. 1958. Wiederfänge markierter Elbaale von der Küste Dänemarks. Arch. Fisch Wiss. 9: 200–202.

McCleave, J.D. and G.P. Arnold. 1999. Movements of yellow- and silver-phase European eels (*Anguilla anguilla* L.) tracked in the western North Sea. ICES J. Mar. Sci. 56: 510–536.

Metcalfe, J.D., M.C. Fulcher, S.R. Clarke, M.J. Challiss, and S. Heatherington. 2009. An archival tag for monitoring key behaviors (feeding and spawning) in fish. In: J.L. Nielsen, H. Arrizabalaga, N. Fragoso, A. Hobday, M. Lutcavage and J. Sibert [eds.]. Tagging and Tracking of Marine Animals with Electronic Devices. Methods and Technologies in Fish Biology and Fisheries 9: 243–254.

Mondini. 1783. De Anguilla Ovariis. De Bononiensis Scientarium et Artium Institute atque Academia Commentarii. Tomus VI. pp. 406–418.

Moore, A. and W.D. Riley. 2009. Magnetic particles associated with the lateral line of the European eel *Anguilla anguilla*. J. Fish Biol. 74: 1629–1634.

Määr, A. 1947. Über die aalwanderung im Baltischen meer auf grund der wanderaalmarkirungsversuche im finnischen und livischen meerbusen in den jahren

1937–1939. Medd. Statens Unders. Försöksanst. Sötvattensfisk, Drottningholm 27: 1–56.

Määr, A. 1949. The migration speed of the silver eel (*Anguilla vulgaris* L.) in the Baltic. Excerptum Apophoreta Tartuensia, Stockholm 1949 pp. 421–427.

Nishi, T., G. Kawamura, and K. Matsumoto. 2004. Magnetic sense in the Japanese eel, *Anguilla Japonica*, as determined by conditioning and electrocardiography. J Exp. Biol. 207: 2965–2970.

Nordqvist, O. 1903. Ålfiskeförsök och ålundersökningar i södra Finland. Fiskeritidskrift för Finland 13: 73–84.

Økland, F., E.B. Thorstad, H. Westerberg, K. Aarestrup and J.D. Metcalfe. 2013. Development and testing of attachment methods for pop-up satellite archival transmitters in European eel. Animal Biotelemetry 1: 3

Palstra, A.P., D.F.M. Heppener, V.J.T. van Ginneken, C. Székely, and G.E.E.J.M. van den Thillart. 2007. Swimming performance of silver eels is severely impaired by the swim-bladder parasite *Anguillicola crassus*. J. Exp. Mar. Biol. Ecol. 352: 244–256.

Petersen, J.C.C. 1905. Larval eels (*Leplocephalus brevirostris*) of the Atlantic coasts of Europe. Meddel. Kommissionen for Havundersogelser, ser. Fiskeri, Bind I, Nr. 5.

Rommel Jr., S.A. and J.D. McCleave. 1973a. Sensitivity of American eels (*Anguilla rostrata*) and Atlantic salmon (*Salmo salar*) to weak electric and magnetic fields. J. Fish. Res. Bd. Canada 30: 657–663.

Rommel Jr., S.A. and J.D. Mccleave. 1973b. Prediction of oceanic electric fields in relation to fish migration. J. Du Conseil 35: 27–31.

Schmidt, J. 1922. The breeding places of the eel. Philos. Trans. Roy. Soc. 211: 179–208.

Sjöberg, N.B. and E. Petersson. 2005. Blankålsmärkning—Till hjälp för att förstå blankålens migration i Östersjön. Fiskeriverket informerar, FINFO 2005: 3.

Stasko, A.B. and D.G. Pincock. 1977. Review of underwater biotelemetry, with emphasis on ultrasonic techniques. Journal of the Fisheries Research Board of Canada 34: 1261–1285.

Svedäng, H. and L. Gipperth. 2012. Will regionalisation improve fisheries management in the EU? An analysis of the Swedish eel management plan reflects difficulties. Marine Policy 36: 801–808.

Svärdson, G. 1976. The decline of the Baltic eel population. Rep. Inst. Freshwat. Res. 55: 136–143

Syrski. 1874. Über die Reproduktions-Organe de Aale. Sitzungsb. d. k. Akad. D. Wissensch., Wien LXIX pp. 315–326.

Tesch, F.-W. 1972. Versuche zur telemetrischen verfolgung der laichwanderung von aalen (*Anguilla anguilla*) in der Nordsee. Helgolander wiss. Meeresunters. 23: 165–183.

Tesch, F.-W. 1974a. Speed and direction of silver and yellow eels, *Anguilla anguilla*, released and tracked in the open North Sea. Ber. Dtsch. Wiss. Komm. Meeresforsch. 23: 181–197.

Tesch, F.-W. 1974b. Influence of geomagnetism and salinity on the directional choice of eels. Helgoland Mar. Res. 26: 382–395.

Tesch, F.-W. 1975. Migratory behavior of displaced homing yellow eels (*Anguilla anguilla*) in the North Sea. Helgol. Wiss. Meeresunters. 27: 190–198.

Tesch, F.-W. 1978. Telemetric observations on the spawning migration of the eel (*Anguilla anguilla*) west of the European continental shelf. Environ. Biol. Fish. 3: 203–209.

Tesch, F.-W. 1979. Tracking of silver eels (*Anguilla anguilla* L.) in different shelf areas of the northeast Atlantic. Rapp. P.-v. Réun. Cons. Int Explor. Mer 174: 104–114.

Tesch, F.-W. 1989. Changes in swimming depth and direction of silver eels (*Anguilla anguilla* L.) from the continental shelf to the deep sea. Aquatic Living Resources 2: 9–20.

Tesch, F.-W. 2003. The Eel (5th edition, J.E. Thorp Ed.). Oxford: Blackwell Science.

Tesch, F.-W., R. Kracht, M. Schoth, D.G. Smith, and G. Wegner. 1979. Report on the eel expedition of FRV Anton Dohrn and R.K. Friedrich Heincke to the Sargasso Sea 1979. ICES CM1979/M:6.

Tesch, F.-W. and A. Lelek. 1973. Directional behavior of transplanted stationary and migratory forms of the eel, *Anguilla anguilla*, in a circular tank. Nether. J. Sea Res. 7: 46–52.

Tesch, F.-W., H. Westerberg, and L. Karlsson. 1991. Tracking studies on migrating silver eels in the Central Baltic. Meeresforschung/Rep. Mar. Res. 35: 193–196.

Tesch, F.-W., T. Wendt, and L. Karlsson. 1992. Influence of geomagnetism on the activity and orientation of the eel, *Anguilla anguilla* (L.), as evident from laboratory experiments. Ecol. Fresh. Fish 1: 52–60.

Trybom, F. 1905. Ålmärkningar i Östersjön 1903 och 1904. Sv. Hydrogr.-Biol. Komm. Skr. II pp. 1–4.

Trybom, F. 1908. Ålmärkningar I Östersjön 1905. Sv. Hydrogr.-Biol. Komm. Skr. III pp. 1–9.

Trybom, F. and G. Schneider. 1908. Die markierungsversuch mit aalen und die wanderungen gekennzeichneter Aale in der Ostsee. Rapp P.-v Réun. Cons. Perm. Int. Explor. Mer 9: 51–59.

Tucker, D.W. 1959. A new solution to the Atlantic eel problem. Nature 183: 495–501.

van Ginneken, V.J.T. and G. van den Thillart. 2000. Eel fat stores are enough to reach the Sargasso. Nature 403: 156–157.

van Ginneken, V., B. Muusze, J. Breteler, D. Jansma, and G. Thillart. 2005a. Microelectronic detection of activity level and magnetic orientation of yellow European eel *Anguilla anguilla* L., in a pond. Environ. Biol. Fish. 72: 313–320.

van Ginneken, V., E. Antonissen, U.K. Muller, R. Booms, E. Eding, J. Verreth, and G. van den Thillart. 2005b. Eel migration to the Sargasso: remarkably high swimming efficiency and low energy costs. J. Exp. Biol. 208: 1329–1335.

van Ginneken, V., S. Dufour, M. Sbaihi, P. Balm, K. Noorlander, M. de Bakker, J. Doornbos, A. Palstra, E.Antonissen, I. Mayer and G. van den Thillart. 2007. Does a 5500-km swim trial stimulate early sexual maturation in the European eel (*Anguilla anguilla* L.)? Comp. Biochem. Physiol. 147A: 1095–1103.

van den Thillart, G., V. van Ginneken, F. Koemer, R. Heijmans, R. van der Linden, and A. Gluvers. 2004. Endurance swimming of European eel. J. Fish Biol. 65: 312–318.

Walton, I. 1653. The complete angler. Available at http://www.gutenberg.org/ebooks/9198

Westerberg, H. 1979. Counter-current orientation in the migration of the European eel. Rapp. P.-v. Réun. Cons. Int Explor. Mer. 174: 134–143.

Westerberg, H. 1984. The orientation of fish and the vertical stratification at fine- and micro-structure scales. In: J.D. McCleave, G.P. Arnold, J.J. Dodson and W.H. Neill [eds.]. Mechanisms of Migration in Fishes, NATO Conf. Ser. IV: Mar. Sci. Vol. 14 pp. 179–203.

Westerberg, H., I. Lagenfelt, and H. Svedäng. 2007. Silver Eel Migration Behavior in the Baltic. ICES Journal of Marine Science: Journal Du Conseil 64: 1457–1462.

Westerberg, H. and I. Lagenfelt. 2008. Sub-sea power cables and the migration behavior of the European eel. Fisheries Management and Ecology 15: 369–375.

Westin, L. 1990. Orientation mechanism in migrating European silver eels (*Anguilla anguilla*): Temperature and olfaction. Mar. Biol. 106: 175–179.

Westin, L. 1998. The spawning migration of European silver eel (*Anguilla anguilla* L.) with particular reference to stocked eel in the Baltic. Fish. Res. 38: 257–270.

Westin, L. 2003. Migration failure in stocked eels *Anguilla anguilla*. Mar. Ecol. 254: 307–311.

Westin, L. and L. Nyman. 1977. Temperature as orientation cue in migrating silver eel (*Anguilla anguilla* L.). Contribution from the Askö Laboratory. No 17.

Westin, L. and L. Nyman. 1979. Activity, orientation, and migration of Baltic eel (*Anguilla anguilla* L.). ICES Journal of Marine Science: Journal Du Conseil 64: 115–123.

Wunsch, C. and R. Ferrari. 2004. Vertical Mixing, Energy, and the General Circulation of the Oceans. Annual Review of Fluid Mechanics 36: 281–314.

CHAPTER 5

Sea Lamprey Migration: A Millenial Journey

Pedro R. Almeida[1,2,*] *and Bernardo R. Quintella*[2,3]

Introduction

The lampreys are a very ancient lineage of vertebrates, with the first recognized fossil found in the Devonian period showing little morphological changes in 360 million years (Gess et al. 2006). Extant lampreys are a small group of 43 species, including anadromous, landlocked, and purely freshwater *taxa* (Renaud 2011; Mateus et al. 2013). Over half the known species are small, non-parasitic, or brook lamprey forms, which never feed during their brief adult lives of six to nine months. The remainder feed as adults in a parasitic manner (Hardisty 1986a).

The sea lamprey *Petromyzon marinus* Linnaeus, 1758, is the largest lamprey, reaching a length of 1.2 m and weighing up to 2.3 kg (Hardisty 1986b). The word *petromyzon* is derived from the Greek meaning stone sucker in reference to the characteristic behavior of adults, during the spawning migration, of attaching to stones with their oral disk to rest during the upstream movement and nest building phases. The word *marinus* is a reference to the environment from which mature adults migrate from after a

[1]Departamento de Biologia, Escola de Ciências e Tecnologia, Universidade de Évora, Largo dos Colegiais, 7004-516 Évora, Portugal.
Email: pmra@uevora.pt
[2]Centro de Oceanografia, Faculdade de Ciências, Universidade de Lisboa, Campo Grande, 1749-016 Lisboa, Portugal.
[3]Departamento de Biologia Animal, Faculdade de Ciências, Universidade de Lisboa, Campo Grande, 1749-016 Lisboa, Portugal.
Email: bsquintella@fc.ul.pt
*Corresponding author

1–2 year period of parasitic feeding in a marine environment, before entering rivers to spawn. The sea lamprey anadromous form is widely distributed on both sides of the North Atlantic, and a smaller form is landlocked in the Great Lakes Basin of North America (Fig. 1). The landlocked sea lamprey is considered non-native and a pest, causing tremendous damage to native fish stocks and the expenditure of large amounts of money in their control (Smith and Tibbles 1980).

The sea lamprey life cycle, and of lampreys in general, can be divided into two completely distinct phases: an adult marine phase and a freshwater larval phase (Fig. 2). The larval phase starts immediately after fecundation, with the embryonic and proammocoete stages. After the absorption of the yolk, the young ammocoetes, approximately 7 mm in length, emerge from the sand of the nest three weeks after the completion of spawning, and are carried downstream to be deposited by the slackening current in areas of fine substrate (Applegate 1950; Potter 1980). The word ammocoete derives from the Greek, meaning "sleeping in sand". Lampreys are regarded as a highly successful group of vertebrates and much of this success is attributed to the protracted freshwater larval phase when, for periods of several years, the ammocoete lies burrowed in fine sediment deposits of rivers and streams, filtering from the water the micro-organisms and organic particles on which it feeds (Hardisty and Potter 1971a). The duration of the sea lamprey larval phase can vary greatly between geographic regions with different climatic regimes (Beamish and Potter 1975; Beamish 1980; Morkert et al. 1998). Larval phase duration is, in part, the result of the time needed to attain a critical size and gather the necessary energetic reserves to initiate metamorphosis (Youson 1980). After a period of 2–8 years in freshwater (Beamish and Potter 1975; Morkert et al. 1998; Quintella et al. 2003), depending on location and environmental conditions, the larva undergoes a metamorphosis. This change in form usually occurs as a consequence of a variety of developmental processes taking place in the organism. In some cases, like for sea lamprey and other anadromous lampreys, metamorphosis is a requirement to prepare organisms for a life in a new habitat, the marine environment (Youson 1980). The term 'transformer' is normally applied to those animals in which the more obvious external changes are still taking place, while the term 'macrophthalmia' or juvenile is used to describe the phase immediately after the completion of metamorphosis when animals are fully transformed. During this final phase, lampreys bear a general resemblance to the adult form and the term macrophthalmia refers to the relatively large size of the eye, which is characteristic of the parasitic species (Hardisty and Potter 1971b). This stage may be said to end with the downstream migration, and the onset of feeding when the animal may be regarded as a young adult.

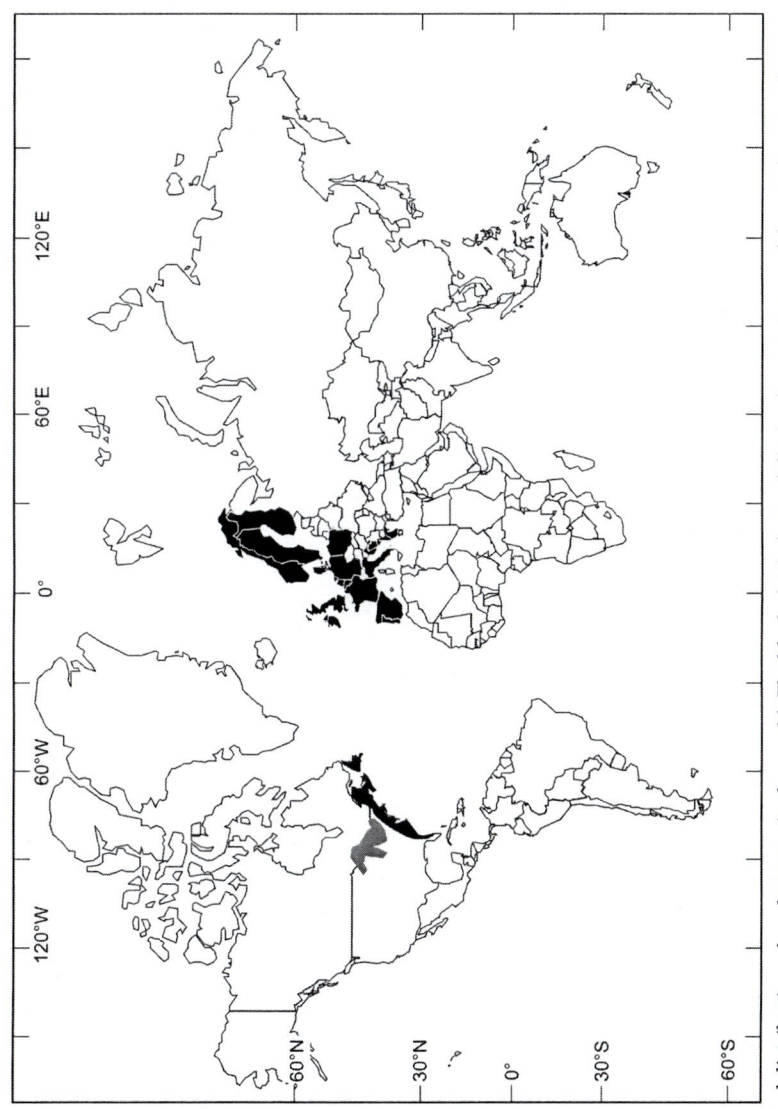

Fig. 1. Geographical distribution of sea lamprey in the world. The black shaded area delimits the occurrence of the anadromous form; the grey area, the landlocked form only found on the North American Great Lakes (map drawn according to the information gathered by: Hubbs and Potter 1971; Beamish 1980; Maitland 1980; Hardisty 1986b; Halliday 1991; Dempson and Porter 1993; Economidis et al. 1999; Holčík et al. 2004; Renaud 2011).

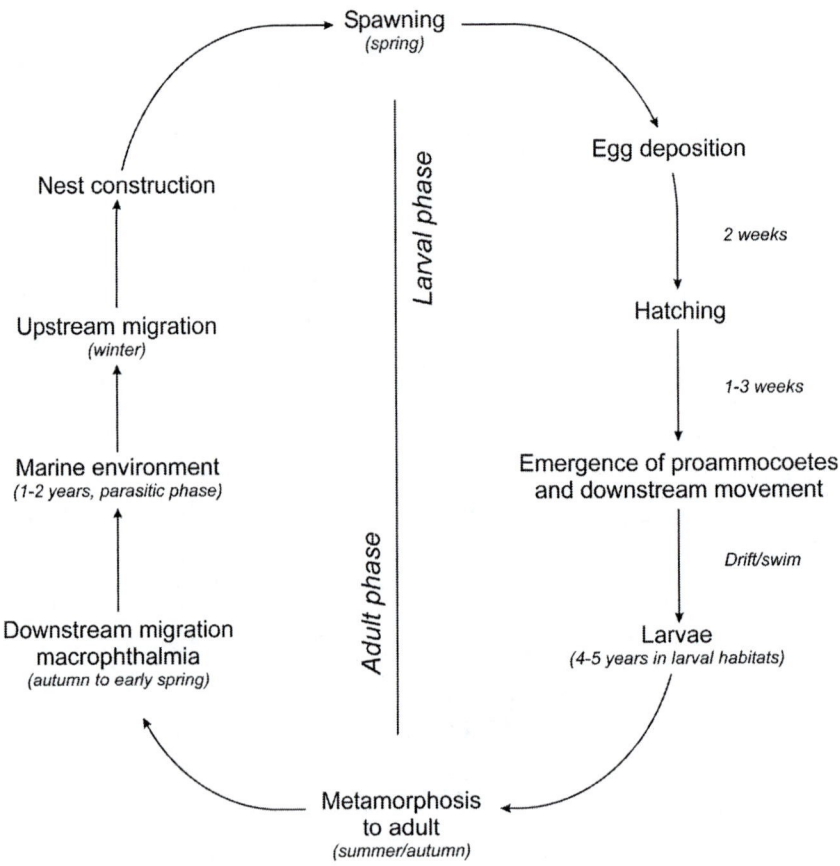

Fig. 2. The anadromous life cycle of the sea lamprey.

The marine period (parasitic phase) of sea lamprey was reported to last from 23 to 28 months (Beamish 1980) but, recently, researchers from the University of Santiago de Compostela, tagged and released sea lamprey macrophthalmia, and one of the tagged individuals was recaptured one year later as entering a river to spawn (Silva et al. 2013a). Little information is available on the feeding ecology of sea lampreys in the marine environment because few specimens have been captured in the ocean and reports of fish with attachment scars are relatively scarce (Farmer 1980). Marine organisms reported to have been preyed upon by sea lamprey include bony fish, elasmobranches, and cetaceans (Beamish 1980; Halliday 1991; Nichols and Hamilton 2004). After this parasitic feeding in the marine environment, sea lamprey initiates a spawning migration to continental waters where it reproduces in the upstream stretches of rivers (Hardisty and Potter 1971b).

Because of its decline across Europe, the sea lamprey was given some legal protection (Lelek 1987; Renaud 1997; Mateus et al. 2012); it is listed under Annex II of the European Union Habitats Directive, the Appendix III of the Bern Convention and is listed in the OSPAR convention list (Convention for the Protection of the Marine Environment of the North-East Atlantic) of threatened and/or declining species. Aquatic pollution, and habitat fragmentation and reduction by construction of large dams, weirs, and other man-made barriers are among the main threats to lamprey populations (Mateus et al. 2012). Since the late 20th century, some European rivers have shown a recent increase in sea lamprey population size following the improvement of water quality (Beaulaton et al. 2008). It is presently assessed as *Least Concern* according to the European Red List of Freshwater Fishes (Freyhof and Brooks 2011).

In many marine animals, especially where the adult is a sedentary filter feeder, the larval stage is an active mobile animal, whose biological function is to spread the population over as large an area as possible, in the hope that at least a few individuals will succeed in reaching habitats suitable for adult survival. The situation in a lamprey is reversed, it is the more active adult that plays the role of exploring the environment, while the filter-feeding ammocoete is relatively passive and sedentary (Hardisty 2006). This chapter aims to review the work done to better understand two demanding episodes of the sea lamprey's life cycle—the downstream (or trophic) migration towards the sea and the upstream (or spawning) migration to the upper stretches of rivers.

1. The Downstream Migration

The metamorphosis precedes the downstream trophic migration of the sea lamprey towards the sea. The ammocoete, although admirably suited to a sedentary and burrowed existence, would be totally incapable of coping with the active life of the adult (Hardisty 2006), namely to be able to accomplish the harsh migrations that are necessary to complete the life cycle. The more obvious morphological changes associated with the metamorphosis of the ammocoete to the adult (i.e., development of the oral disk, the appearance of teeth, eruption of the eyes, enlargement of the fins, and changes in pigmentation (Hardisty and Potter 1971b)) are a prerequisite that enable the lamprey to radically change its mode of life from a filter-feeding larva that lives a sedentary existence in the freshwater environment, to a carnivorous (hematophagous) diet which can be sustained only by a predatory and active mode of life in the marine environment (Hardisty 2006).

1.1. Timing and triggering

The most complete documentation of the downstream migration of the landlocked sea lamprey comes from the studies of Applegate (Applegate 1950; Applegate 1961, and Applegate and Brynildson 1952). According to these authors, the downstream movements begin in the autumn (late October or early November) and continues throughout the winter until the following April. The peak of the downstream migratory activity occurs during late March and early April with a smaller peak in November (Applegate and Brynildson 1952). The amount of water passing downstream has been found to influence the timing of the downstream movement but no relationship was found between changes in water temperature and migratory activity (Applegate and Brynildson 1952).

Data on the downstream migration of the anadromous sea lamprey also exists for the Atlantic drainages of North America and Europe. Observations in the Atlantic coastal rivers of North America have shown that movement in this region takes place in the late autumn and early winter (Davis 1967; Beamish 1980). A similar period has been described for the anadromous populations in European basins (Table 1). A long term monitoring (1997–2010) of the downstream migration of macrophthalmia in a river basin in the Galicia region (Northwest of Spain) recorded movements between October and May (Silva et al. 2013b). According to these authors observations, most of the macrophthalmia (~80%) were captured between January and April with a peak of activity (~30%) during the month of March and sporadic catches in September and June. In France, the Taverny and Elie (2009) study corroborates the downstream migration period for anadromous European populations of sea lamprey between October and March.

A markedly bimodal nature of the timing of the downstream migration was found in North American sea lamprey (landlocked and anadromous form entering Canadian Atlantic Rivers) occurring in both the fall and spring (Applegate 1950; Hardisty 2006). According to Potter (1980) a plausible hypothesis for this phenomenon can be constructed by invoking environmental conditions and the degree of development of the animal. Hardisty (2006) suggests that of the various organs concerned with salt

Table 1. Anadromous sea lamprey migratory periods for the downstream trophic migration of the macrophthalmia and the upstream spawning migration of the adults in Portuguese river basins. The color graduation correlates with the number of individuals engage in the movement for both phases of the life cycle.

Migration	J	F	M	A	M	J	J	A	S	O	N	D
Downstream												
Upstream												

water balance, including the specialist salt pumps located in the gills, it is the achievement of a complete passage through the newly developing foregut that is the major factor influencing the timing of migration and the ability to adapt to a saline environment. Only when this has been completed will macrophthalmia be able to avoid dehydration by swallowing sea water. Thus, if an animal has fully metamorphosed in the fall before temperatures have dropped to minimum values, and is exposed to the stimulus of increased flow, it is likely to start migrating downstream (Youson and Potter 1979). If however, the animal has not completed metamorphosis by the time 'winter' conditions have developed, or has not received a sufficiently strong environmental stimulus to move, it will generally have to wait until the following spring before rising temperatures and discharge rates provide favorable conditions for migration. The downstream migration of sea lampreys at two entirely different times of the year may be of considerable value to this species. The animals migrating in the fall are at an advantage over those that migrate in the spring through the possession of larger lipid reserves and the opportunity to feed earlier. However, if the fall migrants are exposed to deleterious effects during the subsequent winter months, the population of young feeding adults can subsequently be boosted by an influx of fully metamorphosed lampreys in the following spring (Potter 1980).

This bimodal distribution, typical for the North American sea lamprey, with one peak in the autumn and another in spring is not followed by the downstream migration pattern of the European anadromous form. In Europe, the typical migratory period displays a unimodal distribution with a progressive increase in the number of individuals moving, which usually peaks in March, although there are important annual variations (Silva et al. 2013b). Distinct environmental conditions seem to be responsible for these divergent distributions in the numbers of macrophthalmia initiating the downstream movement throughout the migratory period. Climatic conditions in North America (i.e., onset of the winter freeze-up and the break-up of the ice in the following spring caused by rising temperatures and inevitably leading to high water levels) are such as to encourage this separation of autumn and spring migrations (Hardisty 2006). In Western Europe, the milder weather with higher water temperatures is probably responsible for enabling a more continuous and gradual downstream migration (Silva et al. 2013b).

It is clear from some of the studies cited above that the main trigger for stimulating the downstream migration is a marked increase in freshwater discharge (Applegate and Brynildson 1952; Hardisty and Potter 1971b; Potter 1980; Bird et al. 1994; Hardisty 2006). Late fall rains which increase the flow bring down the initial surge of newly transformed individuals. Flood conditions resulting from mid-winter thaws and rains are often accompanied by sudden increases in downstream movement. The greatest downstream migration occurs on the rise and crest of the floods resulting from the general spring break-up in late March or early April (Applegate and Brynildson 1952). For this reason, the peaks of activity may vary from year to year and from one watershed to another (Applegate and Brynildson 1952; Silva et al. 2013b).

1.2. Macrophthalmia migratory behavior

The downstream migration is nocturnal, since macrophthalmia belonging to distinct lamprey species, including the sea lamprey, are invariably caught during darkness (Applegate 1950; Potter and Huggins 1973; Potter 1980). During daylight the outmigrants either burrow or move into protected areas that provide cover.

In terms of habitat preferences, although during the first few weeks of metamorphosis, lamprey transformers are relatively more sedentary than ammocoetes (Quintella et al. 2005), they then move out of slow flowing areas into regions where the substrate is coarser and water flow is faster (Applegate 1950; Potter 1970; Potter and Huggins 1973; Potter 1980). Such trends are also exhibited by the sea lamprey juveniles in Portuguese rivers where they are frequently captured lying hidden during the day among vegetation or the interstitial spaces provided by coarse substrates (unpublished data). According to Hardisty (2006), this change in the choice of habitats between the ammocoete and the macrophthalmia stage are almost certainly linked to the reorganized respiratory systems and higher oxygen requirements of the latter stage. The behavior shown by the migrants, in terms of habitat preferences, makes it easier to understand why the onset of the downstream movement is associated with high water levels. The transformed animals, having abandoned their previous larval habitats known as 'ammocoete beds' (areas where current velocity is usually slow, and the accumulation of soft silt and sand provide a suitable substrate for the burrowing activities of ammocoetes (Almeida and Quintella 2002)) for areas where the bottom consists of coarser materials and faster water currents, will be well positioned to detect any surges in current that may herald the onset of flooding (Hardisty 2006).

One of the most striking characteristics of the migration of newly transformed sea lampreys is the suddenness with which large numbers of individuals leave the river bed and move downstream. To quote Applegate and Brynildson (1952) "Under the impetus of rising waters, a virtual emergence takes place and large numbers of the small lampreys travel downstream on the rise and crest of the floodwaters, this surge frequently ending as suddenly as it began". The same authors also suggested that the sea lamprey downstream migration is passive. This view was based on the observations that lampreys rarely attempted to accelerate their downstream movement, that their movements were casual, and that individuals were occasionally seen drifting tail-first.

There is no data for juvenile sea lamprey swimming capacities but findings for the outmigrating Pacific lamprey *Entosphenus tridentatus* (Richardson 1836) show that their swimming performance is low compared to other anguilliform teleosts (Dauble et al. 2006). Burst speed of juvenile *E. tridentatus* was on average 5.2 body lengths s^{-1} (71 cm s^{-1}), with a sustained swim speed of 23 cm s^{-1} and a critical swimming speed of 36 cm s^{-1} (2.4 BL s^{-1}). McCleave (1980) reported burst speeds of elvers of American eel, *Anguilla rostrata* (Lesueur 1817) up to 7.5 BL s^{-1}. If we compare the swimming performances between the two anguilliform species, the elvers swimming capacity is ca. 50% higher than juvenile lamprey. The weak swimming ability of juvenile lamprey suggests they could be challenged by a lengthy migration to the oceanic feeding grounds.

1.3. Onset of feeding in continental waters

The metamorphosis of the sea lamprey generally begins in the summer and ends during the autumn season (Hardisty and Potter 1971b; Quintella et al. 2003). Some of the macrophthalmia seem to migrate to sea at the end of the metamorphosis but there are others that only move to sea several months latter (Potter and Beamish 1977). In European sea lamprey populations, the bulk of the migrants seem to leave continental waters during the late winter/early spring period (Bird et al. 1994; Silva et al. 2013b). The period between the final transformations associated with metamorphosis (October-November) and the downstream migration to the sea can go up to an average 3–4 months in European rivers (Silva et al. 2013b).

Several studies suggest the existence of hematophagous feeding in continental waters for the anadromous sea lamprey (Table 2). The work of Davis published in 1967 provided evidence that a population of young adult sea lampreys were feeding on landlocked Atlantic salmon—*Salmo salar* (Linnaeus 1758)—in a coastal lake in Maine, and for this reason had presumably not left fresh water. Small sea lamprey adults of variable size (167–351 mm) attack alewife *Alosa pseudoharengus* (Wilson 1811) and spawning-run Atlantic salmon at locations up to 120 km from the estuary (Beamish and Potter 1975). The data indicate that some individuals start feeding in the winter while others do not commence parasitism until the late spring, and that this species does not always migrate downstream to saline conditions immediately after metamorphosis. Occasionally, sea lampreys were also seen attached to American shad *Alosa sapidissima* (Wilson 1811) and white suckers *Catostomus commersonii* (Lacepède 1803)

Table 2. Record of marine organisms parasitized by sea lamprey with reference to the ecology of the host species and the environment (sea/estuary/river) where the attachment was observed.

Host taxon	Environment	Ecology	Reference
CHONDRICHTHYES			
Basking sharks (*Cetorhinus maximus* Gunnerus, 1765)	sea	Pelagic-oceanic; oceanodromous	Beamish (1980)
Sandbar shark (*Carcharhinus plumbeus* Nardo, 1827)	sea	demersal; oceanodromous	Jensen & Schwartz (1994)
Dusky sharks (*Carcharhinus obscurus* Lesueur, 1818)	sea	demersal; oceanodromous	Jensen & Schwartz (1994)
Tiger sharks (*Galeocerdo cuvier* Péron & Lesueur, 1822)	sea	demersal; oceanodromous	Jensen et al. (1998)
Greenland shark (*Somniosus microcephalus* Bloch & Schneider, 1801)	sea	demersal	Gallant et al. (2006)
OSTEICHTHYES			
Twaite shad (*Alosa fallax* Lacepède, 1803)	river	pelagic-neritic; anadromous	Silva et al. (2013b)
American shad (*Alosa sapidissima* Wilson, 1811)	river	pelagic-neritic; anadromous	Potter and Beamish (1977)/ Beamish (1980)
Alewife (*Alosa pseudoharengus* Wilson, 1811)	river	pelagic-neritic; anadromous	Beamish and Potter 1975/ Beamish (1980)
Blueback shad (*Alosa aestivalis* Mitchill, 1814)	river	pelagic-neritic; anadromous	Beamish (1980)
Atlantic menhaden (*Brevoortia tyrannus* Latrobe, 1802)	estuary	pelagic-neritic; oceanodromous	Mansuetti (1962)
Golden grey mullet (*Liza aurata* Risso, 1810)	estuary	pelagic-neritic	Silva et al. (2013b)
Atlantic cod (*Gadus morhua* L., 1758)	sea	demersal; oceanodromous	Beamish (1980)
Atlantic herring (*Clupea harengus*, L., 1758)	sea	demersal; oceanodromous	Farmer (1980)
Haddock (*Melanogrammus aeglefinus* L., 1758)	sea	demersal; oceanodromous	Beamish (1980)
Saithe (*Pollachius virens* L., 1758)	sea	demersal; oceanodromous	Beamish (1980)
Red hake (*Urophycis chuss* Walbaum, 1792)	sea	demersal; oceanodromous	Farmer (1980)
Atlantic mackerel (*Scomber scombrus* L., 1758)	sea	pelagic-neritic; oceanodromous	Farmer (1980)
Atlantic salmon (*Salmo salar* L., 1758)	river/sea	demersal; anadromous	Beamish and Potter (1975)/ Beamish (1980); Silva et al. (2013b)

Table 2. contd....

Table 2. contd.

Host taxon	Environment	Ecology	Reference
Sea trout (*Salmo trutta* L., 1758)	river	pelagic-neritic; anadromous	Silva et al. (2013b)
Brook trout (*Salvelinus fontinalis* Mitchill, 1814)	river	demersal; anadromous	Beamish (1980)
Swordfish (*Xiphias gladius* L., 1758)	sea	pelagic-oceanic; oceanodromous	Beamish (1980)
Striped bass (*Morone saxatilis* Walbaum, 1792)	—	demersal; anadromous	Beamish (1980)
Bluefish (*Pomatomus saltatrix* L., 1758)	—	pelagic-oceanic; oceanodromous	Beamish (1980)
Squeteague (*Cynoscion regalis* Bloch and Schneider, 1801)	estuary	demersal; oceanodromous	Beamish (1980)
CETACEA			
Fin whales (*Balaenoptera physalus* L., 1758)	sea	Pelagic-oceanic	Japha (1910)
Sei whales (*Balaenoptera borealis* Lesson, 1828)	sea	Pelagic-oceanic	Japha (1910)
Harbour porpoise (*Phocoena phocoena* L., 1758)	sea	Pelagic-oceanic	van Utrecht (1959)
Western North Atlantic right whales, (*Eubalaena glacialis* Müller, 1776)	sea	Pelagic-oceanic	Nichols and Hamilton (2004)
Minke whales (*Balaenoptera acutorostrata* Lacépède, 1804)	sea	Pelagic-oceanic	Nichols and Tscherter (2011)
Killer whales (*Orcinus orca* L., 1758)	sea	Pelagic-oceanic	Samarra et al. (2012)

which were present in much lower numbers, in the lake-like extensions of the St. John River system (New Brunswick) during May (Potter and Beamish 1977). Mansuetti (1962) suggested that in the autumn, the recently transformed anadromous sea lamprey underwent a downstream migration to an estuarine environment where, during the subsequent winter months, they fed on Atlantic menhaden, *Brevoortia tyrannus* (Latrobe 1802). In the spring, many of these young adults attached themselves to anadromous clupeids that were moving back into the rivers and in this way re-entered fresh water.

A recent study by Silva et al. (2013b) confirmed this onset of the parasitic feeding in fresh water in European rivers by observing small juvenile sea lamprey attached to resident brown trout *Salmo trutta* (Linnaeus, 1758) in a Spanish watershed, corroborating observations by Davis (1967) and Potter and Beamish (1977) for North American populations. According to this study, 10%–30% of the macrophthalmia start their hematophagous feeding in the river before the downstream migration, with the remaining 70%–90% of the individuals starting in the estuary. In this brackish environment, a particularly important prey species was identified—the golden grey mullet *Liza aurata* (Risso 1810), which is very abundant in European estuaries (Silva et al. 2013b). Mugilidae species that perform movements between estuarine environment and adjacent marine coastal areas, may constitute a key element to the life cycle of the European sea lamprey. On European Atlantic coasts, mullets spend their reproduction period off-shore, on the continental shelf. For the *L. aurata*, spawning occurs from August to October, for the catadromous thin-lipped grey mullet *Liza ramada* (Risso 1827) spawning migration occurs between November and February, and for the thick-lipped grey mullet *Chelon labrosus* (Risso 1827) spawning takes place between December and February (Almeida 1996). In the recent study by Silva et al. (2013b) large amounts of juvenile sea lampreys were observed feeding on *L. aurata* from November to May. It is possible that the sea lamprey may select an intermediary prey in the estuary that, due to their migratory movements, guarantees a 'ride' with a precise timing to the continental shelf. After arriving at the continental shelf, the mullets' spawning areas, the young adult sea lamprey will probably attack a different type of prey. In open sea, it is possible that the host species selected is distinct in terms of ecological habits and size. The potential host must be large enough to support the attachment of fast growing sea lampreys.

2. The Upstream Migration

Following the completion of their marine trophic phase, feeding primarily on blood and muscle tissue of fish during an approximately 1–2 year period (Beamish 1980), adult sea lampreys re-enter freshwater and migrate to upstream river stretches where they build nests, spawn and die (Larsen 1980).

2.1. Timing, triggering and distance covered

The spawning migration can be divided into three stages: (i) migration from the ocean or estuary into rivers or streams; (ii) pre-spawning holding in brackish or fresh water; and (iii) upstream movement within rivers and streams to spawning sites (Clemens et al. 2010). The passage from sea water to fresh water habitats is a particular stressful stage of the migration, and sea lamprey use estuaries as an acclimation chamber, where they shift from a saline basis osmoregulation process to a fresh water orientated one. Riverine entry requires excretion of large volumes of urine, cessation of drinking, intestinal atrophy, and a reversal in ion transport across the gills to allow survival in fresh water (Bartels and Potter 2007). This justifies the pre-spawning holding stage mentioned before, the duration of which is unknown, in the lower part of the rivers where fishing pressure is usually high, making this an extra constraint to their migration. It is also common to observe lampreys in upstream reaches that maintained their position for several weeks, before undergoing a new movement (Almeida et al. 2002a).

The timing and extent of the sea lamprey's spawning migration varies throughout its geographical range. In the east coast of North America, it ranges from September to March (Beamish 1980). In Portuguese rivers, spawning migration begins in December and peaks between February and March (Table 1; Machado-Cruz et al. 1990; Alexandrino 1990; Oliveira et al. 2004), with spawning occurring between April and June (Almeida et al. 2000). In Britain (Severn River), sea lamprey migration begins in February and continues through May and June, while spawning occurs between the end of May and early July (Hardisty 1986b; M. Lucas, personal communication). It is assumed that the species parasitic feeding mode, together with the anadromous behavior, often makes the adults range great distances from the spawning areas. Like other diadromous species, sea lampreys developed a series of physiological and behavior solutions to overcome the environmental constraints encountered during their migration. Due to their anadromous character, sea lampreys are obliged to spawn in fresh water.

The marine phase of their life cycle might influence the timing of the spawning migration, namely the physical-chemical characteristics of the feeding grounds, and/or the targeted hosts. Unpublished data from the Portuguese coast shows the existence of two hypothetic groups with distinct diets. These two distinct diet profiles were observed in animals caught in the main Portuguese river basins. One was typical of a top predator of a marine food web with a planktonic support, while the other was much more diverse, including the same planktonic markers, together with biochemical clues that probably resulted from a parasitic phase that targeted fish consumers of detritus and benthic algae.

The upstream spawning migration is triggered by flow variations and temperature. The increased migratory activity observed during periods of higher river discharge is, probably, a behavior adopted by sea lampreys to overcome difficult passage stretches, enabling the migrants to reach upstream spawning sites (Almeida et al. 2002a; Andrade et al. 2007; Binder et al. 2010). Increased river discharge at night (i.e., hydropeaking) has proven to stimulate lamprey movements, although a reduction in upstream progression, in terms of ground speed, is also observed (Almeida et al. 2002a).

The migration distance depends on the size of the river, the location of suitable spawning areas, and the length of the river stretches downstream impassable barriers (Hardisty 1986b). In the Iberian Peninsula, an estimated 80% loss of accessible habitat for the sea lamprey was caused by the obstruction of the lower stretches of all major rivers. The historical available habitat for the sea lamprey in the rivers Douro, Tagus, Guadiana, Guadalquivir, and Ebro was, respectively, 516 km, 633 km, 648 km, 394 km, and 680 km (Mateus et al. 2012). In Portugal, the present distribution of the sea lamprey is quite limited, with spawning areas located between 27 km (Cávado River) and 150 km (Tagus River) from the mouth of the estuary, but since all the major rivers used by this species present an impassable dam, these obstructions represent the actual upstream limit of their migration (Almeida et al. 2002b; Mateus et al. 2012). In Britain, distances of sea lamprey spawning areas are 10 km–100 km from the tidal limit (Hardisty 1986b; M. Lucas, personal communication).

In the Delaware and Susquehanna River systems (North America), sea lampreys are known to migrate 320 km upstream (Bigelow and Schroeder 1948). Although Daniels (2001) questions the 725 km required for the sea lampreys to get to the Lake Ontario through the Saint Lawrence River (North America), there are evidences that in the past, the longest European runs could reach 850 km in the Rhine River (Hardisty 1986b).

2.2. Fueling migration

The overwhelming majority of migrants rely primarily on fat to fuel their migratory journeys. Consequently, before setting off on their journeys, migrants can store enormous amounts of fat (Dingle 1996). In fish, and depending on the species, triacylglycerols are stored in various tissues including muscle, liver, sub-dermal tissue and mesenteries (Adams 1999), and lampreys store large amounts of triacylglycerols in liver and body wall muscles (Plisetskaya 1980). Lamprey liver and somatic muscles are capable of lipid oxidation and fatty acids are used as oxidative substrates. Triacylglycerols are stored fundamentally as a long-term source of energy that can be used particularly when the energy requirements of the animal are very high like during the migration process that precedes reproduction (Tocher 2003). The neutral lipid contents of anadromous sea lamprey sampled in the beginning of the upstream spawning migration ranged in the muscle between 22% and 29% of the total muscle dry weight against 7%–8% in the liver (Lança et al. 2011).

In addition to the morphological and physiological changes undertaken to prepare for freshwater life, anadromous lampreys also stop feeding. The environmental or physiological cues that stimulate lamprey to detach from their host and embark on the free-swimming, non-trophic spawning migration are unknown; however, during this migration, lampreys survive solely on endogenous energy reserves. As a consequence of this natural starvation, lamprey must mobilize lipid and protein reserves accumulated during the parasitic phase, and experience dramatic and well-documented changes in body morphology (Larsen 1980; Lança et al. 2011). Sea lampreys at the beginning of spawning migration presented a normal pattern for marine fish lipids, where monounsaturated and saturated fatty acids were the most representative class of fatty acid. However, muscle fatty acid profile was not correlated with liver fatty acid which means that sea lamprey muscle acts as a fat depot site and probably is responsible for most of the fatty acid oxidation during spawning migration (Lança et al. 2011).

2.3. Adult migratory behavior

Sea lampreys do not show evidences of homing; adults use olfactory clues to select the optimum spawning/nursery areas following an innate recognition of optimal habitats (Vrieze et al. 2010; Vrieze et al. 2011). For a poor swimmer this is a strategy that makes sense. Nevertheless, we must bear in mind that most of the evidence of lack of philopatry came from trapping records in the Laurentian Great Lakes, which showed that adult sea lamprey are highly selective in the choice of spawning streams (Morman et al. 1980), choosing only those streams that had high densities of larval lamprey (Moore and Schleen 1980). Tagging of out-migrant juvenile sea lamprey and recapture of them as adults on the spawning grounds also provided evidence for a lack of homing to natal streams (Bergstedt and Seelye 1995). In fact, evidence of lack of homing obtained from spawning migrants returning from the Atlantic are based on genetic studies (Bryan et al. 2005; Waldman et al. 2008; unpublished data), but on the other hand, recent studies on heart muscle fatty acid profiles and morphometric using Portuguese sea lampreys show evidence of a restricted dispersion in the ocean, leading to a certain degree of geographical fidelity and stock structure (unpublished data). There is also evidence of absence of exchange among sea lamprey populations spawning in the west and south-east Atlantic coasts (Rodríguez-Muñoz et al. 2004; unpublished data).

Lampreys are negatively phototaxic, moving upstream in freshwater primarily during dusk and darkness (Almeida et al. 2000; Almeida et al. 2002a) and seeking refuge before dawn (Andrade et al. 2007). Similar nocturnal behavior was observed by Almeida et al. (2000) during the estuarine phase of sea lamprey migration in the Mondego River basin (Portugal). According to Andrade et al. (2007), sea lampreys are more likely to search for shelter in rivers when the cumulative distance of their migration exceeds 6.4 km. The adaptive value of nocturnal behavior might be related to the greater protection afforded by darkness.

Lampreys are poor swimmers (in terms of propeller efficiency) compared to many other fishes, mainly because they swim using the anguilliform mode of locomotion, where the body is thrown into undulations with each undulation pushing against the water (Webb 1978). However, when migrating over long distances, anguilliform swimmers may swim four to six times more efficiently (rate of useful energy expenditure divided by the total rate of energy consumption) than non-eel-like fishes (van Ginneken et al. 2005). When swimming through slow-flow river stretches, adult sea lampreys are capable of maintaining a constant pattern of activity, corresponding to an average ground speed of 0.76 body lengths s^{-1} (Quintella et al. 2009), although the most typical swimming records point out to a ground speed between 0.2–0.4 body lengths s^{-1} (Andrade et al. 2007).

Many non-anguilliform species assume a 'burst-and-glide' gait, rather than continuous swimming, to enhance their locomotory performance under high velocities. Similarly, lamprey can use their oral disc to attach to substrate and rest in between bouts of energetic swimming, a strategy referred to as 'burst-and-attach' (Quintella et al. 2009). In areas of fast water velocity, a combination of intermittent burst swimming and periods of rest when attached to the substrate has been recorded as characteristic to lamprey behavior (Applegate 1950; Hardisty and Potter 1971b; Haro and Kynard 1997; Mesa et al. 2003; Quintella et al. 2004). After each move forward, the sea lamprey rests, but each forward thrust requires a longer rest period, as found in the study by Quintella et al. (2009), where each gain resulting from a burst movement of c. 67 s was followed by a rest period of c. 99 s before another attempt was made to move further. This highly active swimming is the most inefficient form of activity with respect to energy costs (Beamish 1978) and can only be achieved for short periods of time. Nevertheless, the absence of swim bladders to sustain neutral buoyancy (Hardisty and Potter 1971b) and the less efficient anguilliform propulsion used by lampreys (Webb 1978; van Ginneken et al. 2005) makes this burst-and-attach pattern the most suitable to overcome rapid flow reaches or man-made obstacles in terms of performance, and is probably the most energetically conservative (Quintella et al. 2004).

2.4. Obstacle passage

To overcome natural barriers, some lamprey species have developed remarkable abilities. Sea lampreys use the oral disc to maintain station in high flows (Quintella et al. 2009). This strategy helps the animals to move through areas of difficult passage, but can only be utilized if adequate attachment surface on the substrate is available (Quintella et al. 2004). The Adams and Reinhardt (2008) study refers to an adequate substrate for a successful lamprey attachment, which should be a regular surface constructed of a slightly rugous material allowing the fimbriae and oral disk to form a seal.

The decline of lamprey populations has been attributed to multiple factors, including commercial fishing (Almeida et al. 2002b; Masters et al. 2006), pollution (Renaud 1997), and loss of or reduced access to key habitats due to river engineering (Renaud 1997; Nunn et al. 2008; Oliveira et al. 2004; Lucas et al. 2009; Almeida et al. 2002b). Even though it is generally accepted that adult anadromous lampreys must have access to upstream spawning areas, these species are rarely considered during the design or modification of traditional fish passage structures (Kemp et al. 2011; Moser et al. 2011). Consequently, most fish passage facilities are not efficient for lamprey (Bochechas 1995; Haro and Kynard 1997; Laine et al. 1998; Lucas and Baras 2001; Moser et al. 2002). This is due to both the relatively poor swimming performance of the lamprey (Beamish 1974; Quintella et al. 2004) and its unique use of the oral disc to attach and rest in high velocity situations (Haro and Kynard 1997; Quintella et al. 2004). According to Quintella et al. (2009), vertical slot and pool and weir type fishways are favorable for successful adult sea lamprey passage. Small, near vertical barriers, rock ramp fishways, and nature-like bypass structures are also likely to be highly successful for passing adult sea lampreys (M. Lucas, personal communication).

Sea lamprey also appears to be obstructed by turbulence in fishways (Haro and Kynard 1997). In a study performed on the Garonne and Dordogne rivers (France) it was documented that pool-type fish passage facilities with vertical slots and fish elevators were successfully used by sea lamprey during their upstream migration, although the efficiency of each type of installation was never estimated (Travade et al. 1998). According to Bochechas the Borland-type fish pass installed in 1986 in the Belver Dam (Tagus River, Portugal) frequently fails to work, and has proved inefficient for anadromous species, including the sea lamprey (Bochecas 1995). In contrast, the Holyoke Dam fish lift on the Connecticut River (at river kilometer 140) regularly passes sea lamprey upstream (Stier and Kynard 1986). In some cases, dam removal may be the only recourse (Gardner et al. 2012).

3. Adaptative Advantages of a Successful Life Cycle

The benefit that migratory lampreys obtain from moving between two environments to complete the life cycle is identical to other anadromous species. Substantial osmotic, bioenergetic and predation-exposure costs in moving between freshwater and oceanic ecosystems certainly occur, but anadromous fish benefit from the generally reduced predation on early life stages in fresh waters and access to the greater trophic resources in marine environments (Gross 1987). For the sea lamprey in particular, the richness of the sea resources for its diet is measured not only in terms of numbers of potential host species/individuals but also the necessary size of the parasitized species to sustain an adult lamprey that can reach a considerable size.

Lampreys have an unusual life cycle for vertebrates since they possess a considerably longer larval stage when compared with adulthood. The sea lamprey displays a larval stage that lasts approximately 4–5 years (Quintella et al. 2003), whereas the parasitic adult only remains in the marine environment for a period no longer than 1–2 years (Beamish 1980; Silva et al. 2013a). The long larval stage, which for the majority of the anadromous fish would be a disadvantage due to the increased risk of predation during this vulnerable period, is more beneficial than detrimental to lampreys since the larva spend that period of the life cycle burrowed in river sediments and more or less sedentary. If on the one hand, the microphagous diet of the ammocoete is energetically poor and responsible for a delayed growth during the consequently long larval stage (Hardisty and Potter 1971a; Moore and Mallat 1980), on the other, the inexhaustibility of the food source (i.e., particulate organic matter, algae and other microscopic organisms (Sutton and Bowen 1994)) is an obvious advantage, since it facilitates the sedentary mode of life that maximizes the probability of survival.

Ammocoetes are highly dependent on the habitat characteristics of rivers to be able to accumulate sufficient lipid reserves to sustain the nontrophic metamorphosis process and the consequent downstream migration toward the feeding grounds in marine environment. The most important habitat feature is the type of substrate (sand and silty sediments are usually the preferred), or variables indirectly related with this physical characteristic, which somehow reflect the substrate granulometry, like the river gradient or current velocity. Taking this in account, it is important that the adults, when returning from their parasitic trophic period, select wisely the system to spread their descendents. Evidence points out to the absence of homing in the sea lamprey (Bergstedt and Seelye 1995; Waldman et al. 2008), a behavior that would allow the return to the natal river, to habitats of established spawning success for the adults and favorable for ammocoete rearing. Waldman et al. (2008) suggest that the sea lamprey does not home; rather it exhibits regional panmixia while using a novel 'suitable river' strategy to complete its life cycle. These findings are sustained by genetic analysis and ancillary information on pheromonal communication. The same authors suggest that this panmixia is one outcome of an alternative life-history strategy from that employed by other anadromous fishes, and directly related with its parasitism. The spectrum of potential hosts in marine waters is extensive and includes pelagic oceanodromous fishes that perform long distance migrations (see Table 2). An adult lamprey that parasitizes a highly mobile species engaged in long scale movements may disperse widely in ocean waters and increase the risk of not returning to continental waters to reproduce. Is this a good adaptive strategy? We believe so, but only for a small part of the population. We will come back to this discussion at the end of this concluding section.

Pheromones are one of the keys to the success of the anadromous lampreys' life-history. The sea lamprey produces at least two types of chemoattractants. The stream-dwelling larva release a migratory bile acid-based pheromone (Bjerselius et al. 2000) that acts as a stimuli to the adult sea lampreys, guiding them to watersheds that give some guarantee of past success for both spawning grounds, as well as nursery areas for larval growth. Mature, spermiated, male sea lampreys also release a potent sex pheromone that induces preference and searching behavior in ovulated female lampreys ascending to the upstream spawning areas (Li et al. 2002). This chemical cue to migration, based on recognition of the presence of conspecifics, may constitute a successful alternative strategy to homing to species that may disperse widely in the marine environment if they happened to attach to a highly mobile host (Waldman et al. 2008).

Juvenile lampreys generally exhibit protracted outmigration timing. Sea lamprey may move downstream immediately after metamorphosis or remain in continental waters for a period up to four months (Silva et al. 2013b). One possible selective advantage of having a highly variable timing for the downstream migration is that it might help to increase the chances that at least some members of the population will migrate and start feeding under favorable circumstances (Youson and Potter 1979). Young feeding adults that are already at sea can subsequently be boosted by an influx of fully metamorphosed lampreys in the following spring (Potter 1980). According to Potter (1980), the animals that initiate the downstream movement in the beginning of the trophic migration period (autumn) are at an advantage over those that migrate in the spring through the possession of larger lipid reserves and the opportunity to feed earlier. However, this is not necessarily true. Several authors confirmed the onset of the sea lamprey parasitic feeding in fresh or estuarine waters (Mansueti 1962; Beamish and Potter 1975; Potter and Beamish 1977; Silva et al. 2013b). Those individuals that initiate the parasitic feeding in continental waters can grow up to ca. 30–40 cm in length (Beamish 1980; Silva et al. 2013b) and are obviously capable of better migratory performances due to their increased size and body condition (Beamish 1974). However, it is curious to notice that the outmigrating peak for the European sea lamprey overlaps the spawning migration period of some eurhyhaline species that move from the estuaries and adjacent coastal waters to the continental shelf for reproductive purposes. We hypothesize here that one of the advantages of the onset of the parasitic feeding in continental waters, with the consequent retarded movement to the sea, may be related with the ecology of the already identified important host species in European estuaries, the mullets

(Mugilidae) (Silva et al. 2013b; unpublished data). As regards the North American sea lamprey population, anadromous salmonids and clupeids were identified as common hosts for the onset of parasitic feeding of the sea lamprey in continental waters (Table 2; Mansueti 1962; Beamish and Potter 1975; Potter and Beamish 1977). Small adult lampreys have restricted swimming capacities (Dauble et al. 2006). Consequently, a 'ride' on an intermediary host species that are moving to the sea, to their spawning (euryhaline or catadromous hosts) or feeding grounds (anadromous hosts), is certainly a factor that would decrease the mortality rate during the initial parasitic feeding in the predator-rich marine environment.

The fate of the juvenile lamprey in the oceanic environment is linked to the choice of the hosts during the parasitic stage of the life-cycle. The range of dispersal of sea lamprey individuals is largely dependent on the mobility of the fish of attachment. Sea lamprey individuals that preferentially parasitize pelagic hosts might be engaged in a life history strategy that enhances long distance passive dispersion but also increases the risk of not returning to continental waters to reproduce and complete their life cycle. A more conservative strategy, with animals feeding mainly on demersal hosts with limited mobility and associated to the continental slope, or even to the ocean basin floor, may favor the successful return to rivers with adequate conditions for adult spawning activities and ammocoete rearing. Rodríguez-Muñoz et al.'s (2004) study showed an absence of genetic exchange among sea lamprey populations spawning in the west and east Atlantic coasts. Apparently, the lamprey's transoceanic dispersion is limited, taking into account the apparent reproductive isolation detected with the genetic analysis done so far (Rodríguez-Muñoz et al. 2004; Bryan et al. 2005). The absence of differences between haplotypic frequencies among collections within western (Waldman et al. 2008) and eastern Atlantic rivers (unpublished data), points out to a regional panmixia and assumes reproductively isolated populations—the European and North American sea lamprey populations. Dispersal is often density-dependent in a wide variety of taxa (Amarasekare 2004). Due to population density, dispersal may relieve pressure for resources in an ecosystem, and competition for these resources may be a selection factor for dispersal mechanisms (Irwini and Taylor 2001). A study that identified morphological differentiation among sea lamprey groups entering western Iberian rivers may reveal distinct ecological strategies triggered by different density levels during the continental larval phase (unpublished data). Lampreys originally from more populated river basins may have a higher tendency to engage in a more dispersal directed behavior during the adult oceanic phase by attacking, preferentially, pelagic fish and thus being responsible for a wider dispersion of the species throughout the neighboring marine areas and, consequently, river basins. The fatty acid signature of sea lamprey muscle detected two feeding strategies used during the species' parasitic marine trophic phase (unpublished data). Adult sea lamprey may parasitize fish with a planktonic support, and these may include fish that feed directly on plankton or piscivorous fish that predate the planktivorous ones; or fish that besides the panktonic markers, show biochemical clues of having fed on fish consumers of detritus and benthic algae. The results of this study seem to be in accordance with the data collected with the morphometric analysis.

Part of the information presented in this section are hypotheses that we have drawn based on the data that we have recently collected (i.e., genetic, morphological, biochemical), and on the studies performed by other researchers also engaged in advancing knowledge about a relatively unknown period of the sea lamprey life cycle, the parasitic phase in the oceanic environment. Further studies are needed to reveal the mysteries of the marine phase of the sea lamprey life cycle. Since we believe that sea bed topography could play a major role in the dispersion of this species and, consequently, in a hypothetic stock structure, it would not come as a surprise if the results from the European Atlantic coast would differ from the ones obtained in the North American coast. This information is particularly urgent in geographical areas where the species presents some conservation problems, since the success of any management plan implemented towards the sustainability of the species' fisheries, depends on sound information regarding the species life cycle.

Acknowledgements

The authors wish to express their thanks to the Japan Society for the Promotion of Science for supporting financially the attendance of P.R. Almeida to the 1st International Conference on Fish Telemetry, 12–18 June 2011, Sapporo, Japan; and also the participation in the public lectures presented during the 25th anniversary of the Akiyama Life Science Foundation, Hokkaido University. This research was partially funded by Fundação para a Ciência e a Tecnologia, COMPETE (Programa Operacional Factores de Competitividade), QREN (Quadro de Referência Estratégico Nacional) and FEDER (Fundo Europeu de Desenvolvimento Regional) through project funding (RECRUIT-PTDC/BIA-BEC/103258/2008), and through the pluriannual funding program to the Center of Oceanography (PEst-OE/MAR/UI0199/2011).

References

Adams, S.M. 1999. Ecological role of lipids in the health and success of fish populations. *In:* M.T. Arts and B.C. Wainman [eds.]. Lipids in freshwater ecosystems. Springer-Verlag, New York, USA pp. 132–160.

Adams, R.D. and U.G. Reinhardt. 2008. Effects of texture on surfaceattachment of spawning-run sea lampreys *Petromyzon marinus:* a quantitative analysis. J. Fish Biol. 73: 1464–1472.

Alexandrino, P.J.B. 1990. Dispositivos de transposição de barragens para peixes migradores, em deslocação para montante. Unpublished MSc thesis, Universidade do Porto, Oporto, Portugal.

Almeida, P.R. 1996. Biologia e Ecologia de *Liza ramada* (Risso, 1826) e *Chelon labrosus* (Risso, 1826) (Pisces, Mugilidae) no Estuário do Mira (Portugal). Inter-relações com o Ecossistema Estuarino. PhD thesis. Universidade de Lisboa, Lisbon, Portugal.

Almeida, P.R. and B.R. Quintella. 2002. Larval habitat of the sea lamprey (*Petromyzon marinus* L.) in the River Mondego (Portugal). *In:* M.J. Collares-Pereira, M.M. Coelho, and I.G. Cowx [eds.]. Freshwater fish conservation: options for the future. Fishing News Books, Blackwell Science, Oxford, UK pp. 121–130.

Almeida, P.R., B.R. Quintella, and N.M. Dias. 2002a. Movement of radio-tagged anadromous sea lamprey during the spawning migration in the River Mondego (Portugal). Hydrobiologia 483: 1–8.

Almeida, P.R., B.R. Quintella, N.M. Dias, and N. Andrade. 2002b. The anadromous sea lamprey in Portugal: biology and conservation perspectives. *In:* M. Moser, J. Bayer, and D. MacKinlay [eds.]. The biology of lampreys, symposium proceedings. International Congress on the Biology of Fish. American Fisheries Society, Vancouver, Canada.

Almeida, P.R., H.T. Silva, and B.R. Quintella. 2000. The migratory behavior of the sea lamprey Petromyzon marinus L., observed by acoustic telemetry in the River Mondego (Portugal). *In:* A. Moore and I. Russel [eds.]. Advances in fish telemetry. CEFAS, Lowestoft, Suffolk, UK.

Amarasekare, P. 2004. The role of density-dependent dispersal in source–sink dynamics. J. Theoret. Biol. 226: 159–168.

Andrade, N.O., B.R. Quintella, J. Ferreira, S. Pinela, I. Póvoa, P. Sílvia, and P.R. Almeida. 2007. Sea lamprey (*Petromyzon marinus* L.) spawning migration in the Vouga river basin (Portugal): poaching impact, preferential resting sites and spawning grounds. Hydrobiologia 582: 121–132.

Applegate, V.C. 1950. Natural history of the sea lamprey, *Petromyzon marinus*, in Michigan. Special Scientific Report 55, United States Department of the Interior, Michigan. USA.

Applegate, V.C. 1961. Downstream movement of lampreys and fish in the Carp Lake River, Michigan. Special Scientific Report-Fisheries 387, U.S. Fish and Wildlife Service. USA.

Applegate, V.C. and C.L. Brynildson. 1952. Downstream movement of recently transformed sea lampreys, *Petromyzon marinus*, in the Carp Lake River, Michigan, Trans. Am. Fish. Soc. 81: 275–290.

Bartels H. and I.C. Potter. 2007. Cellular composition and ultrastructure of the gill epithelium of larval and adult lampreys: implications for osmoregulation in fresh and seawater. J. Exp. Biol. 207: 3447–3462.

Beamish, F.W.H. 1974. Swimming performance of adult sea lamprey, *Petromyzon marinus*, in relation to weight and temperature. Trans. Am. Fish. Soc. 103: 355–358.

Beamish, F.W.H. 1978. Swimming capacity. *In:* Fish physiology, Vol. 7. W.S. Hoar and D.J. Randall [eds.]. Academic Press, New York, USA pp. 101–187.

Beamish, F.W.H. 1980. Biology of the North American anadromous sea lamprey, *Petromyzon marinus*. Can. J. Fish. Aquat. Sci. 37: 1924–1943.

Beamish, F.W.H. and I.C. Potter. 1975. The biology of the anadromous sea lamprey. (*Petromyzon marinus*) in New Brunswick. J. Zool. 177: 57–72.

Beaulaton L., C. Taverny, and G. Castelnaud. 2008. Fishing, abundance and life history traits of the anadromous sea lamprey (Petromyzon marinus) in Europe. Fisher. Res. 92: 90–101.

Bergstedt, R.A. and J.G. Seelye. 1995. Evidence for a lack of homing by sea lampreys. Trans. Am. Fish. Soc. 124: 235–239.

Bigelow, H.B. and W.C. Schroeder. 1948. Fishes of the western North Atlantic, part 1, cyclostomes. Memoir of the Sears Foundation for Marine Research 1: 29–58.

Binder, T.R., R.L. McLaughlin, and D.G. McDonald. 2010. Relative importance of water temperature, water level, and lunar cycle to migratory activity in spawning-phase sea lampreys in Lake Ontario. Trans. Am. Fish. Soc. 139: 700–712.

Bird, D.J., I.C. Potter, M.W. Hardisty, and B.I. Baker. 1994. Morphology, body size and behavior of recently-metamorphosed sea lampreys, *Petromyzon marinus*, from the lower River Severn, and their relevance to the onset of parasitic feeding. J. Fish Biol. 44: 67–74.

Bjerselius, R., W. Li, J.H. Teeter, J.G. Seelye, P.B. Johnsen, P.J. Maniak, G.C. Grant, C.N. Polkinghorne, and P.W. Sorensen. 2000. Direct behavioral evidence that unique bile acids released by larval sea lamprey function as a migratory pheromone. Can. J. Fish. Aquat. Sci. 57: 557–569.

Bochechas, J.H.R. 1995. Condições de funcionamento e de eficácia de eclusas para peixes: casos das barragens de Crestuma-Lever e de Belver. Unpublished MSc thesis, Universidade Técnica de Lisboa, Lisbon, Portugal.

Bryan, M.B., D. Zalinski, K.B. Filcek, S. Libants, W. Li, and K.T. Scribner. 2005. Patterns of invasion and colonization of the sea lamprey (*Petromyzon marinus*) in North America as revealed by microsatellite genotypes. Mol. Ecol. 14: 3757–3773.

Clemens, B.J., T.R Binder, M.F. Docker, M.L. Moser, and S.A. Sower. 2010. Similarities, differences, and unknowns in biology and management of three parasitic lampreys of North America. Fish. 35: 580–596.

Daniels, R.A. 2001. Untested assumptions: the role of canals in the dispersal of sea lamprey, alewife, and other fishes in the eastern United States. Environ. Biol. Fish. 60: 309–329.

Dauble, D.D., R.A. Moursund, and M.D. Bleich, 2006. Swimming behavior of juvenile Pacific lamprey, *Lampetra tridentata*. Env. Biol. Fish. 75: 167–171.

Davis, R.M. 1967. Parasitism by newly transformed anadromous Sea lampreys on landlocked salmon and other fishes in a coastal Maine lake. Trans. Am. Fish. Soc. 96: 11–16.

Dempson J.B. and T.R. Porter. 1993. Occurrence of sea lamprey, Petromyzon marinus, in a Newfoundland river, with additional records from the northwest Atlantic. Can. J. Fish. Aquat. Sci. 50: 1265–1269.

Dingle, H. 1996. Migration: The biology of life on the move. Oxford University Press, New York, USA.

Economidis P.S., A. Kallianiotis and H. Psaltopoulou. 1999. Two records of sea lamprey from the north Aegean Sea. J. Fish Biol. 55: 1114–1118.

Farmer, J.G. 1980. Biology and physiology of feeding in adult lampreys. Can. J. Fish. Aquat. Sci. 37: 1751–1761.

Freyhof, J. and E. Brooks. 2011. European Red List of Freshwater Fishes. Publications Office of the European Union, Luxembourg.

Gallant, J., C. Harvey-Clark, R.A. Myers and M.J.W. Stokesbury. 2006. Sea lamprey attached to a Greenland Shark in the St. Lawrence Estuary, Canada. NE. Naturalist 13: 35–38.

Gardner, C., S.M. Coghlan, and J. Zydlewski. 2012. Distribution and abundance of anadromous sea lamprey spawners in a fragmented stream: current status and potential range expansion following barrier removal NE. Naturalist 19: 99–110.

Gess, R.W., M.I. Coates, and B.S. Rubidge. 2006. A lamprey from the Devonian period of South Africa, Nature 443: 981–984.

van Ginneken, V., E. Antonissen, U.K. Muller, R. Booms, E. Eding, J. Verreth, and G. van den Thillart. 2005. Eel migration to the Sargasso: remarkably high swimming efficiency and low energy costs. J. Exp. Biol. 208: 1329–1335.

Gross, M. 1987. The evolution of diadromy in fishes. Am. Fish. Soc. Symp. 1: 14–25.

Halliday, R.G. 1991. Marine distribution of the sea lamprey (Petromyzon marinus) in the northwest Atlantic. Can. J. Fish. Aquat. Sci. 48: 832–842.

Hardisty, M.W. 1986a. General introduction to lampreys. *In:* J. Holčík [eds.]. The freshwater fishes of Europe Vol.1, Part I—Petromyzontiformes. Aula-Verlag, Wiesbaden pp. 19–83.

Hardisty, M.W. 1986b. *Petromyzon marinus* Linnaeus, 1758. *In:* J. Holčík [eds.]. The freshwater fishes of Europe Vol.1, Part I—Petromyzontiformes. Aula-Verlag, Wiesbaden pp. 94–116.

Hardisty, M.W. 2006. Lampreys: life without jaws. Forrest Text, Ceredigion. UK.

Hardisty, M.W. and I.C. Potter. 1971a. The behavior, ecology and growth of larval lampreys. *In:* M.W. Hardisty and I.C. Potter (eds.). The biology of lampreys, Vol. 1. Academic Press, London. UK pp. 85–125.

Hardisty, M.W. and I.C. Potter. 1971b. The general biology of adult lampreys. *In:* M.W. Hardisty and I.C. Potter (eds.). The biology of lampreys, Vol. 1. Academic Press, London. UK pp. 127–247.

Haro, A. and B. Kynard. 1997. Video evaluation of passage efficiency of American shad and sea lamprey in a modified Ice Harbor fishway. North Am. J. Fish. Manag. 17: 981–987.

Holčík, J., A. Delić, M. Kučinić, V. Bukvić and M. Vater. 2004. Distribution and morphology of the sea lamprey from the Balkan coast of the Adriatic Sea. J. Fish Biol. 64: 514–527.

Hubbs, C.L. and I.C. Potter. 1971. Distribution, phylogeny and taxonomy In: M.W. Hardisty and I.C. Potter (eds.). The biology of lampreys, Vol. 1. Academic Press, London. UK pp. 1–65.

Irwin, A.J. and P.D. Taylor. 2001. Evolution of altruism in a stepping-stone population with overlapping generations. Theor. Popul. Biol. 60: 315–325.

Japha, A. 1910. Weitere Beiträge zur Kenntnis der Walhaut. Zool. Jahrb. 12: 711–718.

Jensen, C., F.J. Schwartz and G. Hopkins. 1998. A sea lamprey (Petromyzon marinus)-tiger shark (Galeocerdo cuvier) parasitic relationship off North Carolina. J. Elisha Mitchell Scient. Soc. 114: 72–73.

Jensen, C. and F.J. Schwartz. 1994. Atlantic Ocean occurrences of the sea lamprey, *Petromyzon marinus* (Petromyzontiformes, Petromyzontidae), parasitizing sand-bar, *Carcharhinus plumbeus*, and dusky, C. obscurus (Carcharhiniformes: Carcharhinidae), sharks off North and South Carolina. Brimleyana 21: 69–72.

Kemp, P.S., I.J. Russon, A.S. Vowles, and M.C. Lucas. 2011. The influence of discharge and temperature on the ability of upstream migrant adult river lamprey (*Lampetra fluviatilis*) to pass experimental overshot and undershot weirs. River Res. Appl. 27: 488–498.

Laine, A., R. Kamula, and J. Hooli. 1998. Fish and lamprey passage in a combined Denil and vertical slot fishway. Fish. Manag. Ecol. 5: 31–44.

Lança, M.J., C. Rosado, M. Machado, R. Ferreira, I. Alves-Pereira, B.R. Quintella, and P.R. Almeida. 2011. Can muscle fatty acid signature be used to distinguish diets during the marine trophic phase of sea lamprey (*Petromyzon marinus*, L.)? Comp. Biochem. Physiol. B. 159: 26–39.

Larsen, L.O. 1980. Physiology of adult lampreys, with special regard to natural starvation, reproduction, and death after spawning. Can. J. Fish. Aquat. Sci. 37: 1762–1779.

Lelek, A. 1987. The Freshwater Fishes of Europe Vol. 9, Threatened Fishes of Europe. Aula-Verlag, Wiesbaden.

Li, W., A.P. Scott, M.J. Siefkes, H. Yan, Q. Liu, S. Yun, and D.A. Gage. 2002. Bile acid secreted by male sea lamprey that acts as a sex pheromone. Science 296: 138–141.

Lucas, M.C. and E. Baras. 2001. Migration of freshwater fishes. Blackwell Science, Oxford, England, UK.

Lucas, M.C., D.H. Bubb, M.H. Jang, K. Ha, and J.E.G. Masters. 2009. Availability of and access to critical habitats in regulated rivers: effects of low-head barriers on threatened lampreys. Freshw. Biol. 54: 621–634.

Machado-Cruz, J.M., A.C.N. Valente, and P.J.B. Alexandrino. 1990. Contribuição para a caracterização ecológica e económica da pesca de migradores a jusante da barragem de Belver, Rio Tejo. Proceedings 1º Congresso do Tejo. Que Tejo, que futuro? 189–200.

Maitland, P.S. 1980. Review of the ecology of lampreys in northern Europe. Can. J. Fish. Aquat. Sci. 37: 1944–1952.

Mansueti, J. 1962. Distribution of small, newly metamorphosed sea lampreys, *Petromyzon marinus*, and their parasitism on menhaden, *Brevoortia tyrannus*, in mid-Chesapeake Bay during winter months. Chesapeake Sci. 3: 137–139.

Masters, J.E.G., M.H. Jang, K. Ha, P.D. Bird, P. Frear, and M.C. Lucas. 2006. The commercial exploitation of a protected anadromous species, the river lamprey (*Lampetra fluviatilis* (L.)), in the tidal River Ouse, north-east England. Aquat. Conser. Mar. Freshw. Ecosyst. 16: 77–92.

Mateus, C.S., R. Rodríguez-Muñoz, B.R. Quintella, M.J. Alves, and P.R. Almeida. 2012. Lampreys of the Iberian Peninsula: distribution, population status and conservation. Endang. Species Res. 16: 183–198.

Mateus, C.S., M.J. Alves, B.R. Quintella and P.R. Almeida. 2013. Three new cryptic species of the lamprey genus Lampetra Bonnaterre, 1788 (Petromyzontiformes: Petromyzontidae) from the Iberian Peninsula. Contributions to Zoology 82: 37–53.

McCleave, J.D. 1980. Swimming performance of European eel [*Anguilla anguilla* (L.)] elvers. J. Fish Biol. 16: 445–452.

Mesa, M.G., J.M. Bayer, and J.G. Seelye. 2003. Swimming performance and physiological responses to exhaustive exercise in radio-tagged and untagged Pacific lampreys. Trans. Am. Fish. Soc. 132: 483–492.

Moore, H.H. and L.P. Schleen. 1980. Changes in spawning runs of sea lamprey (*Petromyzon marinus*) in selected streams of Lake Superior after chemical control. Can. J. Fish. Aquat. Sci. 37: 1851–1860.

Moore, J.W. and J.M. Mallat. 1980. Feeding of larval lamprey. Can. J. Fish. Aquat. Sci. 37: 1658–1664.

Morkert, S.B., W.B. Swink, and J.G. Seelye. 1998. Evidence for early metamorphosis of sea lampreys in the Chippewa River, Michigan. N. Am. J. Fisher. Manag. 18: 966–971.

Morman, R.H., D.W. Cuddy, and P.C. Rugen. 1980. Factors influencing the distribution of sea lamprey (*Petromyzon marinus*) in the Great Lakes. Can. J. Fish. Aquat. Sci. 37: 1811–1826.

Moser, M.L., M.L. Keefer, H.T. Pennington, D.A. Ogden, and J.E. Simonson. 2011. Development of Pacific lamprey fishways at a hydropower dam. Fish. Manag. Ecol. 18: 190–200.

Moser, M.L., P.A. Ocker, L.C. Stuehrenberg, and T.C. Bjornn. 2002. Passage efficiency of adult Pacific lampreys at hydropower dams on the Lower Columbia River, USA. Trans. Am. Fish. Soc. 131: 956–965.

Nichols, O.C. and P.K. Hamilton. 2004. Occurrence of the parasitic sea lamprey, *Petromyzon marinus*, on western North Atlantic right whales, *Eubalaena glacialis*. Env. Biol. Fishes 71: 413–417.

Nichols, O.C. and U.T. Tscherter. 2011. Feeding of sea lampreys *Petromyzon marinus* on minke whales *Balaenoptera acutorostrata* in the St. Lawrence Estuary, Canada. J. Fish Biol. 78: 338–343.

Nunn, A.D., J.P. Harvey, R.A.A. Noble, and I.G. Cowx. 2008. Condition assessment of lamprey populations in the Yorkshire Ouse catchment north-east England, and the potential influence of physical migration barriers. Aquatic. Conserv. Mar. Freshw. Ecosyt. 18: 175–189.

Oliveira, J.M., M.T. Ferreira, A.N. Pinheiro, and J.H. Bochechas. 2004. A simple method for assessing minimum flows in regulated rivers: the case of sea lamprey reproduction. Aquat. Cons. 14: 481–489.

Plisetskaya, E. 1980. Fatty acid levels in blood of cyclostomes and fish. Environ. Biol. Fish. 5: 273–290.

Potter, I.C. and F.W.W. Beamish. 1977. The freshwater biology of adult anadromous sea lampreys *Petromyzon marinus*. J. Zool. Lond. 181: 113–130.

Potter, I.C. and R.J. Huggins. 1973. Observations on the morphology, behavior and salinity tolerance of downstream migrating River lampreys (*Lampetra fluviatilis*). J. Zool. Lond. 169: 365–379.

Potter, I.C. 1970. The life cycles and ecology of Australian lampreys of the genus *Mordacia*. J. Zool. Lond. 161: 487–511.

Potter, I.C. 1980. Ecology of larval and metamorphosing lampreys. Can. J. Fish. Aquat. Sci. 37: 1641–1657.

Quintella, B.R., N.O. Andrade, A. Koed, and P.R. Almeida. 2004. Behavioral patterns of sea lampreys spawning migration during difficult passage areas studied by electromyogram telemetry. J. Fish Biol. 65: 1–12.

Quintella, B.R., I. Povoa, and P.R. Almeida. 2009. Swimming behavior of upriver migrating sea lamprey assessed by electromyogram telemetry. J. Appl. Ichthyol. 25: 46–54.

Quintella, B.R., N.O. Andrade, and P.R. Almeida. 2003. Distribution, larval stage duration and growth of the sea lamprey ammocoetes, Petromyzon marinus L., in a highly modified river basin. Ecol. Fresh. Fish 12: 1–8.

Quintella. B.R., N.O. Andrade, R. Espanhol, and P.R. Almeida. 2005. The use of PIT telemetry to study movements of ammocoetes and metamorphosing sea lampreys in river beds. J. Fish Biol. 66: 97–106.

Renaud, C.B. 1997. Conservation status of Northern Hemisphere lampreys (Petromyzontidae). J. Appl. Ichthy. 13: 143–148.

Renaud, C.B. 2011. Lampreys of the world. An annotated and illustrated catalogue of lamprey species known to date. FAO Species Catalogue for Fishery Purposes. N°. 5. FAO, Rome. Italy.

Rodríguez-Muñoz, R., J.R. Waldman, C. Grunwald, N.K. Roy, and I. Wirgin. 2004. Absence of shared mitochondrial DNA haplotypes between sea lamprey from North American and Spanish rivers. J. Fish Biol. 64: 783–787.

Samarra, F.I.P., A. Fennell, K. Aoki, V.B. Deecke and P.J.O. Miller. 2012. Persistence of skin marks on killer whales (*Orcinus orca*) caused by the parasitic sea lamprey (*Petromyzon marinus*) in Iceland. Mar. Mammal Sci. 28: 395–401.

Silva, S., M.J. Servia, R. Vieira-Lanero, S. Barca and F. Cobo. 2013a. Life cycle of the sea lamprey *Petromyzon marinus*: duration of and growth in the marine life stage. Aquat. Biol. 18: 59–62.

Silva, S., M.J. Servia, R. Vieira-Lanero, and F. Cobo. 2013b. Downstream migration and hematophagous feeding of newly metamorphosed sea lampreys (*Petromyzon marinus* Linnaeus, 1758). Hydrobiologia 700: 277–286.

Smith, B.R. and J.J. Tibbles. 1980. Sea Lamprey (*Petromyzon marinus*) in Lakes Huron, Michigan, and Superior: History of Invasion and Control, 1936–78. Can. J. Fish. Aquat. Sci. 37: 1780–1801.

Stier, K. and B. Kynard. 1986. Movement of sea-run sea lampreys, *Petromyzon marinus*, during the spawning migration in the Connecticut River. Fish. Bull. 84: 749–753.

Sutton, T.M. and S.H. Bowen. 1994. Significance of organic detritus in the diet of larval lampreys in the Great Lakes Basin. Can. J. Fish. Aquat. Sci. 51: 2380–2387.

Taverny, C. and P. Elie. 2009. Bilan des connaissances biologiques et de l'etat des habitats des lamproies migratrices dans le bassin de la Gironde—Propositions d'actions prioritaires. Rapport Final. Etude Cemagref, n° 123 Groupement de Bordeaux, Bourdeaux. France.

Travade, F., M. Larinier, S. Boyer-Bernard, and J. Dartiguelongue. 1998. Performance of four fish pass installations recently built on two rivers in south-west France. *In:* M. Jungwirth, S. Schmutz, and S. Weiss [eds.]. Fish migration and fish bypasses. Fishing News Books, Oxford, UK pp. 146–170.

Tocher, R.D. 2003. Metabolism and functions of lipids and fatty acids in teleost fish. Rev. Fish. Sci. 11: 107–184.

van Utrecht, W.L. 1959. Wounds and scars in the skin of the common porpoise *Phocaena phocaena* (L.). Mammalia 13: 100–122.

Vrieze, L.A., R.A. Bergstedt, and P.W. Sorensen. 2011. Olfactory-mediated stream finding behavior of migratory adult sea lamprey (Petromyzon marinus). Can. J. Fish. Aquat. Sci. 68: 523–533.

Vrieze, L.A., R. Bjerselius, and P.W. Sorensen. 2010. Importance of the olfactory sense to migratory sea lampreys *Petromyzon marinus* seeking riverine spawning habitat. J. Fish. Biol. 76: 949–964.

Waldman, J., C. Grunwald, and I. Wirgin. 2008. Sea lamprey Petromyzon marinus: an exception to the rule of homing in anadromous fishes. Biol. Lett. 4: 659–662.

Webb, P. W. 1978. Hydrodynamics: nonscombroid fish. *In:* Fish physiology, Vol. 7. W.S. Hoar and D. J. Randall [eds.]. Academic Press, New York, USA pp. 189–237.

Youson, J.H. 1980. Morphology and physiology of lamprey metamorphosis. Can. J. Fish. Aquat. Sci. 37: 1687–1710.

Youson, J.H. and I.C. Potter. 1979. A description of the stages of metamorphosis in the anadromous sea lamprey, *Petromyzon marinus* L. Can. J. Zool. 57: 1808–1817.

CHAPTER 6

Migratory Behavior of Adult Pacific Lamprey and Evidence for Effects of Individual Temperament on Migration Rate

Mary L. Moser,[1,a,]* *Matthew L. Keefer,*[2] *Christopher C. Caudill*[2] *and Brian J. Burke*[1]

Introduction

Migratory fishes often exhibit extreme variation in individual behavior, which can result in life history diversity. Telemetry and otolith microchemistry techniques have revealed that some individuals of a given anadromous population exhibit extensive marine migrations, whilst others remain in freshwater and may never even leave the vicinity of their natal areas (Gross 1991; Thorpe et al. 1998; Waples et al. 2004; Miller et al. 2011). Telemetry studies are replete with examples of individuals that perform at levels far outside those predicted by a normal distribution (Aarestrup et al. 2002; Cote et al. 2002; Keefer et al. 2004). However, the mechanisms generating variability in migratory behavior and their fitness consequences are unknown for most species. In this paper, we examined individual variation in lamprey migration rates.

[1]Northwest Fisheries Science Center, National Marine Fisheries Service, 2725 Montlake Boulevard East, Seattle, WA 98112, USA.
[a]Email: mary.moser@noaa.gov
[2]Department of Fish and Wildlife Sciences, College of Natural Resources, University of Idaho, Moscow, Idaho 83844-1136, USA.
*Corresponding author

One outcome of individual variation in lamprey migration behavior is the proliferation of different life history types. Lampreys (Petromyzontiformes) are an ancient group of fish that exhibit great life history variation. In most lamprey genera there are pairs of species that have morphologically indistinguishable larvae but very different juvenile and adult forms (Potter 1980). In most cases, significant genetic differences have not been documented for these pairs (Docker 2009), even though the adults of the two species may exhibit completely different morphologies and life histories (large vs. small, anadromous vs. freshwater resident, parasitic vs. nonparasitic). Even within a species, lampreys can exhibit great plasticity. Anadromous parasitic lampreys often have praecox forms that return to freshwater at a small body size after a shortened marine feeding phase (Kucheryavyi 2007), and these may represent intermediaries in the evolution of nonparasitism (Docker 2009). The sea lamprey (*Petromyzon marinus* Linnaeus) invasion of the Laurentian Great lakes is an infamous example of extreme variation in life history. Anadromous in their native range, the sea lamprey has assumed a highly successful freshwater life history that is exemplified by smaller body size, shorter migration distance, and a truncated parasitic phase (Clemens et al. 2010).

Variation in individual behavior dictates the degree of life history diversity that can occur (Bolnick et al. 2003). For anadromous, parasitic lampreys, migration behaviors or "decisions" include whether to participate in seaward migration, when to initiate and complete the trophic phase, when and where to enter freshwater, the route and speed of the spawning migration, and the location of suitable spawning habitat and mates. Migration decisions made by individuals are drawn from a behavioral repertoire that is likely influenced by some combination of genetic predisposition, physiological cues, and environmental experience (Sih et al. 2004). The persistence of behavioral bias or individual "temperament" over a variety of contexts is known as a behavioral syndrome and such syndromes have been well-documented in some model species (Sih et al. 2004; Bell 2005). Here we explore the possibility that individual lamprey exhibit consistent travel rates during multiple segments of their upstream migration.

In lamprey, timing and rate of the adult migration are affected by environmental variation and fish size, and both have important fitness consequences (Clemens et al. 2009; Keefer et al. 2009a; 2009b). Among anadromous lampreys within the Northern Hemisphere, mean ground speeds recorded during the prespawning migration are remarkably similar (Moser et al. 2013); however, most radiotelemetry studies of adult lamprey migration report high variation in individual migration rates within a population (Stier and Kynard 1986; Kelso and Glova 1993; Jellyman et al. 2002; Lucas et al. 2009; Andrade et al. 2007). Furthermore, when confronted with obstacles to migration, the degree of individual variation can be amplified (Almeida et al. 2002). For example, in an analysis of Pacific lamprey, *Entosphenus tridentatus* (Gairdner), passage at two Columbia River hydropower dams, distributions of passage time had exceedingly long tails, illustrating extreme variation in behavior among individuals (Moser et al. 2005).

Individual lamprey temperament may substantially affect migration rate in concert with the effects of lamprey size and proximate environmental variables. To test this idea, we used a large multiyear radiotelemetry database to examine the migration rates of adult Pacific lamprey through three large reservoirs that presented no obvious obstacles to upstream passage. Given the highly plastic behaviors reported for lamprey at hydroelectric dams and other obstacles, we predicted that individuals would also exhibit wide variation in migration rate through reservoirs. Further, we hypothesized that the individual bias in migration rate would be consistent as individual lamprey traversed successive reservoirs. That is, individuals with relatively rapid migration through the first reservoir would also migrate rapidly through the second reservoir.

We used a mixed-effects model derived from random walk advection-diffusion equations to evaluate both the mean movement of the population (drift) and the spread of individuals from the mean migration rate (dispersal). While diffusion-based models have often been used to describe animal movement (Okubo 1980), their utility for travel time analyses of fish migrations has only recently been demonstrated (Zabel and Anderson 1997; Zabel 2002). We assembled reservoir passage data for adult Pacific lamprey from a long-term radiotelemetry database and used a fixed-effects generalized linear model (GLM) to determine which covariates of lamprey passage (environmental variables and physical attributes such as size and sex) were most influential. We then developed a generalized linear mixed-effects model (GLMM; Zuur et al. 2009) that included a term for individual bias in migration rates through three consecutive Columbia River reservoirs. Finally, we used standard model comparison techniques to compare the mixed and fixed-effects models to determine whether there were significant and predictable behavioral differences among individual lamprey.

1. Model Development

Prespawning adult Pacific lamprey were collected at Bonneville Dam on the Columbia River (rkm 235) during 2000–2002 and 2007–2010 (Fig. 1). Lamprey were captured in traps positioned within the fishway, radiotagged using standard protocols (Moser et al. 2002), and released downstream from the dam. In 2000–2002, transmitters used were either 7.7 g (in air) and 11 mm in diameter or 4.5 g and 9 mm in diameter (Lotek Wireless Inc.,

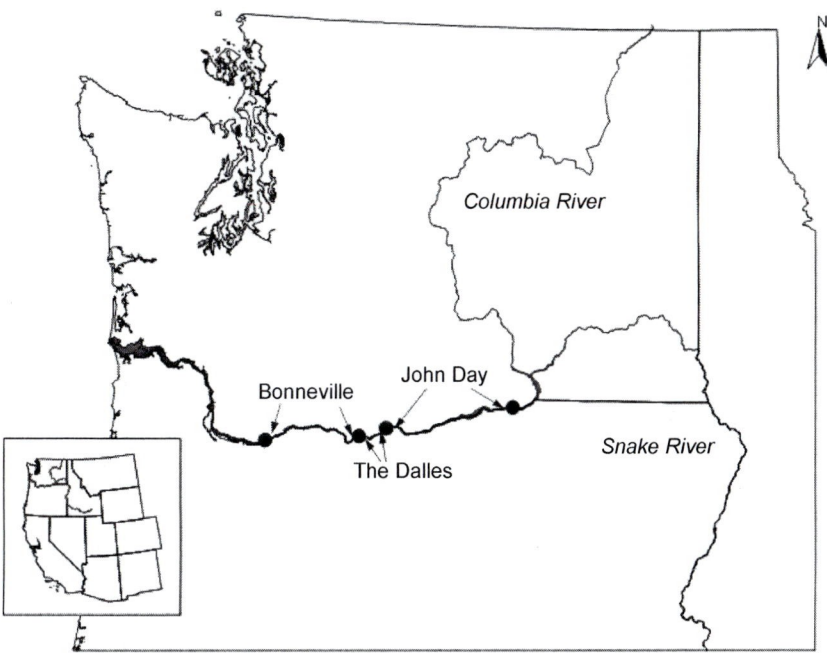

Fig. 1. Study area in the Columbia River drainage. The start and end of the Bonneville, The Dalles, and John Day reservoirs are indicated by arrows.

Newmarket, Ontario),[1] depending on lamprey size. Starting in 2007, when smaller transmitters became available, transmitters used were 2.1 g and 8 mm in diameter (Lotek NTC42L).

Immediately after capture, individual lamprey were anesthetized in a 0.06 ml/l solution of eugenol (the active ingredient in clove oil) and a radio transmitter was implanted into the body cavity. The incision was closed with at least two simple, interrupted sutures and the lamprey were allowed to recover for at least 2 hr prior to release at Columbia rkm 232 (Moser et al. 2002; Johnson et al. 2012). Surgery and handling protocols were reviewed and approved by the University of Idaho Animal Care and Use Committee (ACUC protocol # 200941).

Radiotagged lamprey movements were monitored using an array of fixed-site receivers (Moser et al. 2002; Johnson et al. 2012). Receivers were equipped with digital spectrum processors to receive transmissions on multiple frequencies simultaneously. One or more coaxial cable antennas

---

[1]Use of trade names does not imply endorsement by the National Marine Fisheries Service, NOAA.

were positioned underwater to detect fish as they approached fishway entrances and as the fish exited from a fishway into the forebay of each dam (Fig. 1). Reservoir transit time was defined as time from last detection at a fishway exit to first detection at a fishway entrance of the next upstream dam (Fig. 1). To standardize the transit times across reservoirs, reservoir transit times (in days) were divided by the reservoir length (km) for each individual to obtain the average time an individual took to travel 1 km.

For most travel time estimates, we were able to obtain corresponding data on fish size and sex, as well as date and hour of entry into the reservoir and river discharge and water temperature on that date (covariates). Fish size metrics included length (cm total length), weight (g), and girth at insertion of the first dorsal fin (cm). To minimize tag effects, we tagged only lamprey where transmitter diameter/(π *girth) did not exceed 0.3 (Moser et al. 2007). Lamprey gender was determined by viewing the gonad prior to transmitter insertion, when possible. Mean daily river discharge and water temperature were measured at each dam and reported for the date of lamprey entry into each reservoir (data archived at: http://www.cbr.washington.edu/dart/).

As we were interested in individual-level consistency in behavior, we required at least two observations per fish. Therefore, prior to modeling, individuals without at least two reservoir passage times or with any missing covariate data were removed from the dataset (referred to as the reduced dataset). For each reservoir, the mean migration rates (km/d) for the full dataset and for the reduced dataset were compared using Mann Whitney tests.

Dispersal, which is sometimes referred to as Brownian motion, describes the general spread of individuals from some central point. Dispersal with *drift* is often used in models of animal movement or migration, where drift describes the mean magnitude and direction of movement of a population, and dispersal explains the spread among individuals. Furthermore, one can rewrite the dispersal equation to estimate the time required for an animal to travel a fixed distance, rather than modeling movement rates directly (Fagan 1997; Zabel and Anderson 1997). The resulting equation is referred to as the inverse Gaussian distribution and its probability density function is:

$$f(T \mid L, r, \sigma) = \frac{L}{\sqrt{2\pi\sigma^2 t^3}} e^{-\left(\frac{(tr-L)}{2t\sigma^2}\right)}$$

Where the fitted parameters r and σ are the movement rate (i.e., ground speed) and the spread among individuals, respectively, L is the distance traveled, and t is the travel time.

When fitting this type of model with covariates, one can use an inverse link function, such that the effect of the covariates are inversely related to travel time, and therefore directly and linearly related to ground speed r:

$$r_i = \alpha + \beta_j X_{ij} + \varepsilon_i$$

where the ground speed of individual i is equal to an intercept, α, plus the effect of j covariates, X_{ij} (with coefficients β_j), and residual error ε_i. To determine the magnitude of the effect of an individual in this framework, one could run a series of correlations using the residual travel times, after having accounted for the effects of covariates. However, this would have to be done in a pair wise manner (reservoir by reservoir) and would not allow estimation of individual level residuals among all reservoirs simultaneously. We therefore used a mixed-effects model, which incorporates all reservoirs into a single model while allowing for individual level adjustment to the model intercept. If there was a consistent (across reservoirs), individual specific deviation from the mean travel time, this model formulation would account for it. Since we used an inverse link function, we fit a mean and standard deviation from a normal distribution to the individual deviations from overall mean ground speed, resulting in:

$$r_i = \alpha + \beta_j X_{ij} + b_i + \varepsilon_i$$

where b_i is the individual effect (assumed to be N(0, σ_m) and σ_m is a fitted parameter). An additional benefit of this approach was that the distribution of the individual "random" component, b, could then be directly compared to the magnitude of other model components (such as residual error, ε).

Because we included a large number of potential covariates, and were primarily interested in the individual random effect, we took a simple two step approach to fitting the data. First, we ran a fixed-effects GLM to determine which covariates should be included in the final model. Second, we ran a mixed-effects GLMM, which allowed for an individual random effect for each lamprey, using the set of covariates determined to be important from step one. In both steps, we used an inverse Gaussian distribution and an inverse link function. Potential fixed effect covariates included year, reservoir, day of year (date), water temperature, river discharge, hour of the day of forebay entry (time), and lamprey length, weight, girth, and gender (sex). Year, reservoir, and sex were categorical variables, all others were treated as continuous. We also included a reservoir interaction term for temperature, discharge, date, time, and sex.

We ran the models in a Bayesian framework using JAGS software and the r2jags (Su and Yajima 2011) package in R (R Development Core Team 2011). Many frequentist statistical packages allow for an inverse Gaussian distribution in GLMs; however, the Bayesian approach allows greater flexibility in model formulation, as well as estimates of parameter distribution (as opposed to point estimates). These can be informative tools when analyzing the distribution of individual fish effects. A standard approach to determine whether the individual effect is warranted in the model is to compare two models: one with the effect of individuals and one without. Model comparison techniques are then used to determine which of these is most parsimonious with the data (Zuur et al. 2009). To compare models, we used the deviance information criterion (DIC), which is similar to the Akaike Information Criterion (AIC), penalizing the model for each additional parameter to be estimated (Gelman et al. 2003). All covariates were centered and scaled prior to analysis. For all models, we ran 50,000 Markov chain Monte Carlo iterations with three chains, a burn in period of 25,000, and thinning rate of 25, resulting in 3,000 iterations saved. We assigned uninformative uniform priors on the intercept and individual effect and uninformative normal priors on all covariate parameters.

We evaluated model fit using two posterior predictive checks. For both methods, we sampled replicate data t_{rep} from the posterior parameter distributions. The first check was to determine what percent of the observed travel time data t fell in the 95% credible interval of t_{rep} (Congdon 2010). This is a general check of concordance between observed and predicted values. For the second method, we calculated residuals for each observation as the difference between the mean of the posterior predicted travel time and the observed time. We therefore had a set of observed values, residuals from the fixed-effects model, and residuals from the mixed-effects model. Because each fish had more than one observation, we averaged the model residuals (within each model) to obtain the mean residual per fish. If there were indeed consistently fast and slow fish, these fish-specific model residuals should be reduced when using a random effect of individual in the model (i.e., our mixed-effects model) compared to a similar fixed-effects model.

2. Model Results

The original data set included 913 detections of 657 radiotagged Pacific lamprey. Of these individuals, reservoir passage times were available for 575 (762 observations, referred to as the full dataset). After removing records with missing covariate data, the dataset was further reduced to 673 individual reservoir passage times of 503 fish. Of these, there were 314 passage times of 144 fish that had passed through at least two of the three reservoirs (reduced dataset). This final dataset included 141 passage times through the Bonneville Dam pool, 131 through The Dalles Dam pool, and 42 through the John Day Dam pool. We found no significant difference in migration rate between the full and reduced data sets at The Dalles ($P = 0.61$) and John Day reservoirs ($P = 0.79$), but the mean rate for the full dataset was significantly lower than the mean in the reduced dataset in the Bonneville Reservoir comparison ($P < 0.01$; Fig. 2).

In the reduced dataset, median migration rates were similar among the reservoirs with respective migration rates through Bonneville, The Dalles, and John Day reservoirs of 22.9, 25.7, and 29.6 km/d (Fig. 2). However, we observed an increase in migration rates through each successive reservoir. We also documented substantial variation around the median migration rate, with individual lamprey rates ranging from 3.5 to >67 km/d. As expected, an Inverse Gaussian distribution was appropriate for the data (Fig. 3).

Examination of DIC values indicated that the set of fixed-effect covariates with the best fit to the data included: year, reservoir, time, date, a reservoir×date interaction, discharge, and sex (Fig. 4 and Table 1). Lamprey moved more rapidly through the upstream reservoirs during the later stages of migration. While the time of day when lamprey entered the reservoirs was significant, there were relatively few detections of lamprey movement during the day, preventing a thorough evaluation of time of day. Migration rates were positively correlated with water temperature and negatively correlated with river discharge. Finally, mean migration rates for males were faster than for females (Fig. 4).

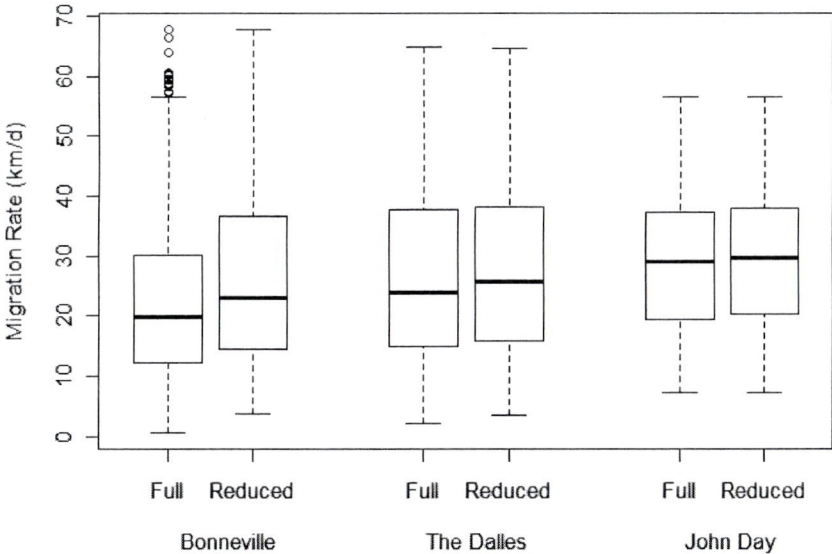

Fig. 2. Box plots of migration rate (km/d) through each reservoir for the full dataset of Pacific lamprey passage times (Full, $n = 762$) and for a subset that included lamprey detected passing through more than one reservoir (Reduced, $n = 314$).

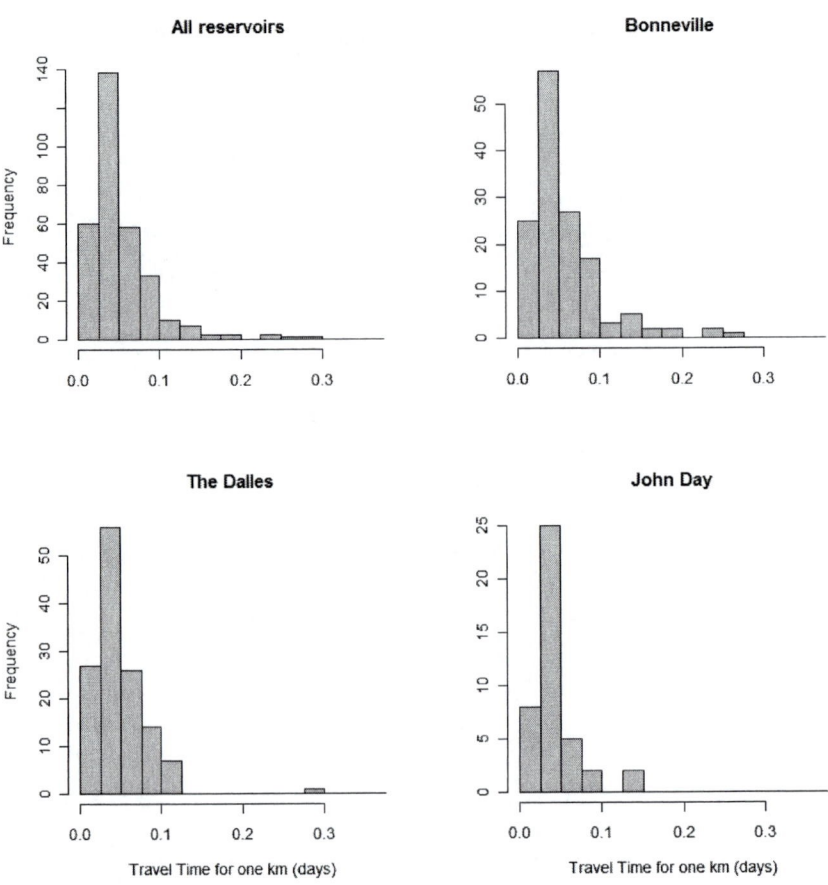

Fig. 3. Frequency distributions of standardized Pacific lamprey travel times through lower Columbia River reservoirs.

The value of DIC for the fixed-effects model (no individual effect) was larger than that of the mixed-effects model with the same predictors by 18.29 units, suggesting that there was a strong individual lamprey effect. Parameter estimates for covariates in the mixed-effects model were lower than for the fixed-effects model in most cases (Table 1). Some lamprey consistently (across reservoirs) traveled up to 5 km/d faster or slower than the overall mean (Fig. 5). However, the proportion of lamprey whose model residuals were either consistently fast or consistently slow was reduced in the mixed-effects model (Table 2), further demonstrating the ability of the mixed-effects model to account for lamprey temperament.

It was also clear that many lamprey did not have much of an offset from the mean. Some of these fish may have been swimming at the mean speed

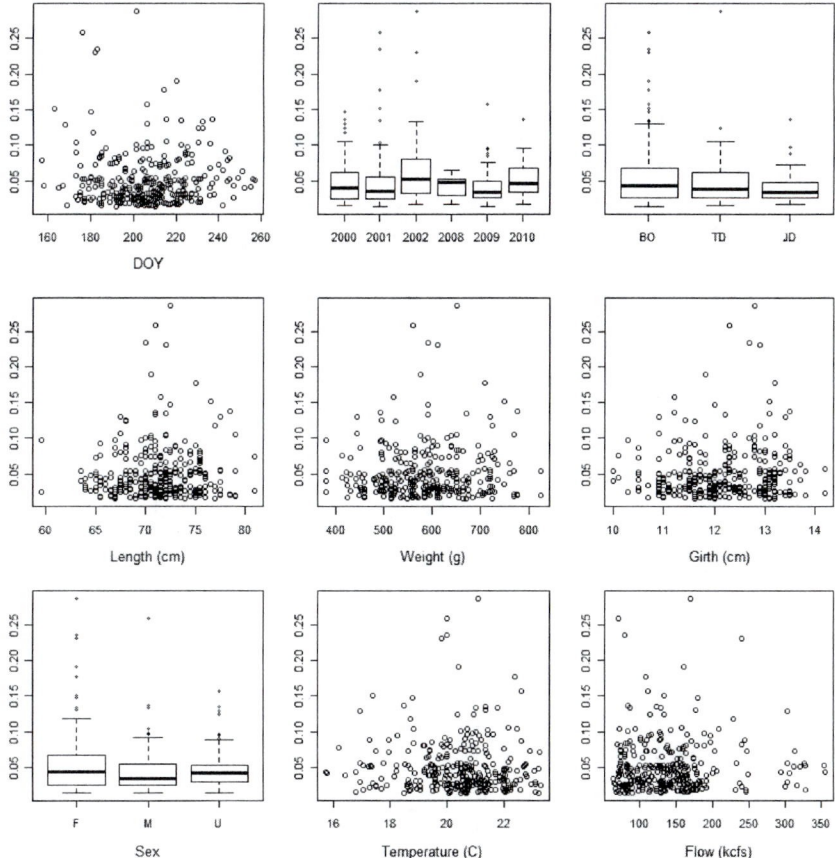

Fig. 4. Simple plots of covariates vs. Pacific lamprey travel times for a 1 km stretch (reduced dataset).

the whole time or they may have been swimming faster in one reservoir and slower in another (which would give them little to no overall offset from the mean).

We evaluated model fit and the magnitude of the individual-based effect using several approaches. For the fixed-effects model, 97.7% of observations fell within the posterior predicted 95% credible interval, suggesting good concordance between observed and predicted values. For the mixed-effects model, this value increased to 99.4% of observations. The fixed-effects model had residual error of 2.68 km/d (Table 1) versus a residual error of 2.45 km/d for the mixed-effects model. This indicates that the addition of individual effects accounted for some variation that would otherwise have been considered random noise (Zabel 2002). Moreover, the estimated standard deviation of the individual effect was 4.90 km/d, which

Table 1. Fixed-effect and mixed-effect model estimates and standard errors for parameters included in the final models. Reservoirs are abbreviated as follows: Bonneville (BO), The Dalles (TD), and John Day (JD). Reference gender was female (M = male, U = unknown).

Parameter	Fixed Effects		Mixed Effects	
	Mean	SD	Mean	SD
Intercept	19.26	1.73	20.67	1.93
Discharge	3.83	1.24	3.32	1.35
Time	2.68	0.72	3.22	0.75
ResJD	6.16	2.76	4.23	2.67
ResTD	1.89	1.49	1.50	1.46
Res×Date-BO	3.76	1.10	3.51	1.21
Res×Date-JD	−5.00	2.65	−4.57	2.68
Res×Date-TD	0.05	1.23	0.03	1.28
Sex-M	2.67	1.69	2.37	1.96
Sex-U	−5.83	2.94	−5.55	3.40
Year 2001	4.14	2.59	4.24	2.78
Year 2002	−8.07	2.13	−7.48	2.42
Year 2008	8.60	4.53	7.76	4.95
Year 2009	8.41	3.27	7.68	3.71
Year 2010	4.18	4.00	4.81	4.45
Residual Error	2.68	0.11	2.45	0.13
Individual Effect	NA	NA	4.90	1.09

was larger than the estimated standard deviation of residuals (2.45 km/d). This and the observation that inclusion of the individual term resulted in a lower residual for nearly all individuals (Fig. 6) further indicated that we accounted for a significant amount of variability with the individual-effect component of the model.

3. Lamprey Temperament and Migration Rate

Adult Pacific lamprey exhibited substantial variation in migration rate through reservoirs. While this modeling exercise revealed covariates that significantly affected lamprey ground speeds, it also provided strong evidence for individual based effects. Based on dispersal theory (Zabel and Anderson 1997), we predicted that lamprey passage times born from a diffusion process with drift would result in an inverse Gaussian distribution. We used that distribution in a GLM framework to estimate

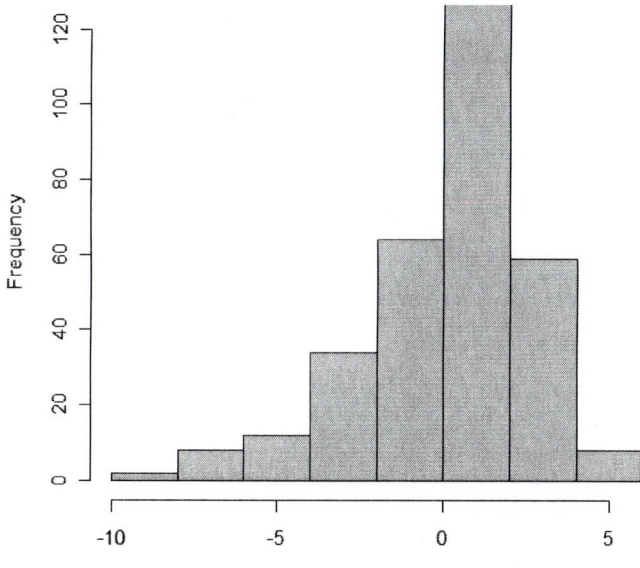

Individual-fish offset to the average ground speed (km/d)

Fig. 5. Distribution of individual Pacific lamprey ground speed offsets from the mean.

Table 2. Percentages of model residuals (fixed-effects and mixed-effects) for adult lamprey that migrated consistently slower (slow-slow) or consistently faster (fast-fast) through Bonneville and The Dalles reservoirs, respectively.

Model	slow-slow	slow-fast	fast-slow	fast-fast
Fixed	14.84	16.41	21.09	47.66
Mixed	9.38	21.88	28.91	39.84

the effects of various covariates on ground speed, and assessed the added impact of individual level variability (i.e., fish temperament) using GLMM. Examination of the model residuals revealed that individual fish passage rates did conform to an inverse Gaussian distribution and that inclusion of individual effects substantially improved model fit.

Mean travel rates through each reservoir ranged from 26–30 km/d, similar to average rates of passage reported in other studies of Pacific lamprey and of other adult anadromous lampreys of the Northern Hemisphere (Moser et al. 2013). In a recent study of acoustically-tagged adult Pacific lamprey, Noyes et al. reported even faster mean migration rates through the Bonneville Reservoir (Noyes et al. 2012). This may have been due to the position of acoustic receivers, which captured lamprey movements without the effect of dam tailraces. In the tailraces of The Dalles and John Day dams, Keefer et al. noted that lamprey became more

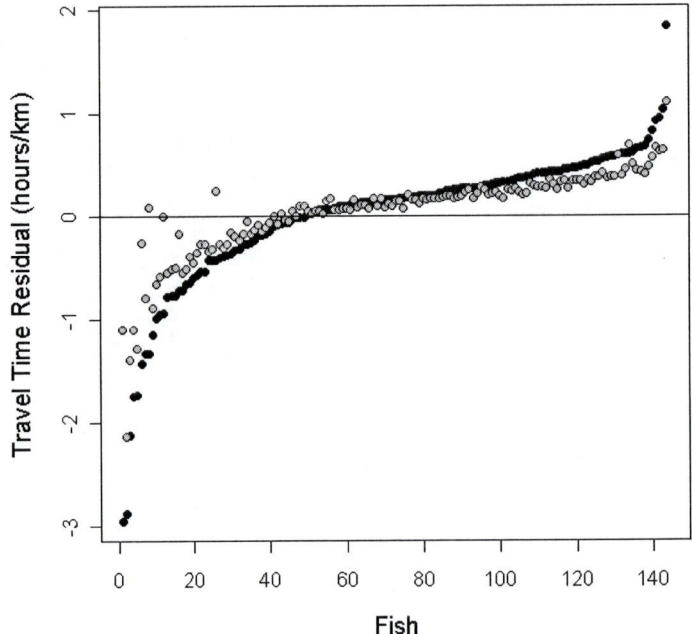

Fig. 6. Model residuals averaged across reservoir observations for each individual for the fixed-effects model (black circles) and mixed-effects model (gray circles). Residuals are ordered for individual Pacific lamprey by the lowest to highest fixed-effect residual. Note that residual values of 0 indicate a perfect fit (horizontal line) and thus residual values that are closer to the line for an individual represent a better fit.

exclusively nocturnal than in the Bonneville Reservoir (Keefer et al. 2012). This would likely result in slower passage through the tailrace portions of the reservoirs (i.e., lamprey switch to moving only at night). Thus, because the tailrace reaches were included in our sample, lamprey passage times through reservoirs were potentially lower than if only the reservoir reaches without tailrace effects were examined. Hence, analysis of the Noyes et al. data for effects of individual temperament may yield an even more pronounced individual effect than was documented here (Noyes et al. 2012).

Collection of individual behavioral observations is labor-intensive, and relatively few studies have the data to support the complicated analyses required to separate individual bias from potential covariates (Bolnick et al. 2003). Telemetry arrays, like the one used in our study, allow collection of large volumes of information on individuals. Consequently, telemetry studies are unique in that individual behaviors are continuously scored, and data on individual activities can be obtained without time-consuming visual observations or introduction of observer bias.

Our analysis suffered from having only two or three sequential comparisons of migration rates. This was because the telemetry array was designed to assess Pacific lamprey passage at dams, rather than their migration through reservoirs. In addition, Pacific lamprey attrition at the Columbia River dams is high (Keefer et al. 2009b), so fewer and fewer lamprey are available for comparisons as they move upstream. However, many telemetry studies employ multiple fixed-site receivers along a migration route (Aarestrup et al. 2002; Almeida et al. 2002; Keefer et al. 2004). In an analysis like ours, the rigor of tests for individual temperament would increase with the number of segments studied, regardless of the number of individuals observed. Data from telemetry studies with multiple migration segments are therefore good candidates for assessment of individual-based effects in other species, even when the number of individuals is relatively low.

We may have underestimated individual effects by necessarily constraining our dataset to only those lamprey with multiple reservoir passage records. In spite of having started with a large initial dataset, our exclusion of fish with missing passage times or covariate data resulted in a considerable reduction of the sample that could be used to test for individual effects. Missing passage times resulted from equipment failures or missed detections, and missing covariates stemmed from gaps in the environmental data stream, which were also due to sporadic equipment failures. Nevertheless, 144 individuals were used in the model, and the migration rates of these individuals did not differ significantly from those of the larger dataset for two of the three reservoirs.

Mean migration rates through Bonneville reservoir were significantly higher for fish in the reduced data set than for those in the full dataset. Adult Pacific lamprey have difficulty negotiating fishways at the dams that separated the reservoirs in this study (Moser et al. 2002; Johnson et al. 2012). Therefore, it is likely that the poorest performers were also the slowest (Caudill et al. 2007), and were winnowed out by the first set of dams. This raises the possibility that individuals having a consistently slow migration rate also have a lower probability of upriver passage.

The fixed effects model indicated that the most important covariate effects were for year, reservoir, discharge, date, time, sex, and the interaction between reservoir and date. While there were interannual differences in migration rate, we could not attribute them to any change in tagging methodology. As in other studies of adult lamprey migration rates (Stier and Kynard 1986; Keefer et al. 2009a), there were clear seasonal effects on migration speed, with lamprey moving more rapidly through the upstream reservoirs later in the migration when water temperature was higher and discharge was lower. Interestingly, there were also significant effects of time of day. Adult lamprey typically exhibit nocturnal activity during the spawning migration (Almeida et al. 2002; Keefer et al. 2012). It is likely that lamprey entering a reservoir at night were active through more of the next 12 hours than those entering at dawn, and this is borne out in our model results.

Not surprisingly, river discharge was negatively correlated with lamprey ground speeds. Lamprey swimming with the same effort would have had slower ground speeds when river discharge was highest, as they would be forced to stem faster currents. Another potential effect of high discharge was the potential dilution of olfactory cues (migratory pheromones produced by larvae) that Pacific lamprey may use to orient (Yun et al. 2011). During periods of high discharge, lamprey may have to search vertically or swim a less direct course in order to find sufficiently high concentrations of olfactory cues for orientation (Vrieze et al. 2011).

While river discharge significantly affected migration rate, and was included in the final model, water temperature and lamprey size did not. In studies of migration timing using lamprey tagged with passive integrated transponders (PIT tags), the largest lamprey migrating during the warmest temperatures traveled fastest and furthest in the Columbia River main stem (Keefer et al. 2009a; Keefer et al. 2009b). It is likely that the relatively small sample size of our radio tagged lamprey ($n = 144$) did not allow detection of temperature and size effects. These effects were strongly significant in studies with large numbers of PIT tagged fish (n = 3,598, Keefer et al. 2009a; Keefer et al. 2009b). Body size variation in our sample was also lower than in Keefer et al. (Keefer et al. 2009a; Keefer et al. 2009b) because the smallest lamprey were not used for radio tagging, to minimize tag effects. In contrast, Pacific lamprey of all sizes were PIT tagged. Moreover, by including only lamprey that had traversed at least two reservoirs, we further reduced body size variation in our sample, as smaller lamprey have a lower probability of passing dams (Keefer et al. 2009b).

While gender was not determined for all lamprey in our sample, sex was still a significant covariate in the final model, with males traveling faster than females. This is a new and interesting finding for Pacific lamprey. In the Laurentian Great Lakes, male sea lamprey precede females on the spawning grounds and emit a sex pheromone to attract females as they near spawning habitats. A similar mechanism is likely at work in Pacific lamprey (Clemens et al. 2010). Our observations suggest that gender specific differences in tributary entry timing may be more related to migration rate than to gender specific differences in migration timing (i.e., males swim faster or are more directed, rather than starting their migration earlier).

Adult Pacific lamprey exhibited great variation in ground speed during reservoir passage, and model comparison techniques revealed that individual bias was a consistent and significant contributor to that variation. Similar or even greater rates of variation in migration speed have been reported in other lamprey studies (Keefer et al. 2009a). However, this is the first analysis to examine whether differences in migration speed can be attributed to individual fish temperament. Some individuals consistently traveled over 5 km/d faster or slower through reservoirs, a relative difference of 15%–25%, than the mean ground speed of the tagged population, even after controlling for environmental variables and the effects of lamprey size and sex. What are potential mechanisms underlying this individual bias in lamprey migration rates? Sih et al. broadly categorized the proximate controls of behavioral syndromes as stemming from genetic differences, environmental experience, and/or neuroendocrine effects (Sih et al. 2004). In the following paragraphs, we consider each of these potential controls.

Based on rangewide studies of mitochondrial DNA and microsatellites, Pacific lamprey are considered to be nearly panmictic (Goodman et al. 2008; Spice et al. 2012). Using amplified fragment length polymorphism, Lin et al. showed some degree of genetic divergence among collections of lamprey from a broad geographic range (Lin et al. 2008); however, they did not specifically connect this divergence to adaptive genetic differences associated with migration. The most recent results from single-nucleotide polymorphism analyses suggest that there may be genetically-controlled adaptations in Pacific lamprey that could manifest in the kinds of individual differences we documented (Hess et al. 2012). Hence, variation in Pacific lamprey migration rate, if genetically based, is not likely to stem from variation among subpopulations (as in philopatric species), but rather from genetic variation within a single population (Pulido 2007).

Environmental experience is also likely to play a role in individual lamprey behavior. Risky individual behaviors, such as the tendency to be active during more hours of the day or to exhibit directed swimming in the water column may be based on past experience in terms of either predator avoidance or successful attachment to a host (Sih et al. 2004). Pacific lamprey, and some other anadromous lamprey species, can enter freshwater over one year prior to spawning and do not feed during their upstream migration (Moser et al. 2013). Therefore, experience from the oceanic trophic phase could play an important role in individual behavior and capacity for activity. Indeed, recent interaction with a host may increase the scope for activity (Claireaux and Lefrancaois 2007) and the physiological ability to swim for longer periods or at a faster rate (Broderson et al. 2008).

Finally, the stability in individual migration rates that we documented is probably also based on individual differences in sensory capability or circulating hormone concentrations. As in other lamprey species, adult Pacific lamprey probably use olfactory cues to orient (Robinson et al. 2009; Yun et al. 2011), and the degree of wandering or holding along reservoir migration routes may be related to individual sensitivity to pheromones.

In adult Pacific lamprey, olfactory sensitivity to migratory pheromones showed the greatest individual variation during the early migratory period in freshwater. This sensitivity could be attributed to individual differences in receptor binding affinity, second messenger function, olfactory receptor cell density, and/or morphology of the olfactory organ (Robinson et al. 2009). Mesa et al. documented little individual variation in circulating sex steroids or thyroid hormones for adult Pacific lamprey during the early phase of spawning migration (Mesa et al. 2010). However, these authors noted that physiological status and tendency to migrate could still be correlated in this fish. Thus it is plausible that the individual effect we detected reflects differences in the "motivational" status among individuals with as yet unmeasured differences in hormone levels.

Extremes in individual behavior often frustrate attempts to categorize or generalize behavioral characteristics. Most fisheries management decisions are based on measures of central tendency rather than dispersion within a population. Moreover, behavioral outliers are usually considered to be random noise around the mean or median. Our results indicated that there can be structure embedded in that noise. By ignoring this structure, important underlying mechanisms may be missed. Fish migrations are replete with examples of individual variation. Rather than eliminating outliers, researchers should embrace and investigate sources of individual variation and potential evidence for behavioral syndromes. Management actions that focus on preservation of mean behaviors (i.e., that protect average performers) may function to limit behavioral variation. This could ultimately constrain life history diversity and the capacity for a population to persist or rebuild (Bolnick et al. 2003).

Acknowledgements

Many people provided field assistance, database management, and administrative support throughout the years of lamprey radiotelemetry research. We thank C. Boggs, T. Bjornn, T. Clabough, W. Daigle, T. Dick, M. Heinrich, B. Ho, M. Jepson, E. Johnson, D. Joosten, S. Lee, M. Morasch, G. Naughton, C. Peery, D. Quaempts, R. Ringe, and K. Tolotti of the University of Idaho. Thanks also to T. Bohn, D. Dey, K. Frick, S. McCarthy, and L. Stuehrenberg of the National Marine Fisheries Service, as well as W. Cavender, D. Ogden, H. Pennington, and J. Roos of the Pacific States Marine Fisheries Commission. For their administrative assistance, we thank D. Clugston, M. Langeslay, T. Mackey, J. Rerecich, and S. Tackley of the U.S. Army Corps of Engineers. E. Buhle provided guidance in modeling techniques and application. J. Butzerin, S. Bourret, J. Hess, M. Kirk, C. Noyes, M. Scheuerell, J. Uber, and R. Zabel provided critical reviews of this manuscript. Funding for this work was provided by the U.S. Army Corps of Engineers.

References

Aarestrup, K., C. Nielsen, and A. Koed. 2002. Net ground speed of downstream migrating radiotagged Atlantic salmon (*Salmo salar* L.) and brown trout (*Salmo trutta* L.) smolts in relation to environmental factors. Hydrobiologia 483: 95–102.

Almeida, P.R., B.R. Quintella, and N.M. Dias. 2002. Movement of radiotagged anadromous sea lamprey during the spawning migration in the River Mondego (Portugal). Hydrobiologia 483: 1–8.

Andrade, N.O., B.R. Quintella, J. Ferreira, S. Pinela, I. Póvoa, P. Sílvia, and P.R. Almeida. 2007. Sea lamprey (*Petromyzon marinus* L.) spawning migration in the Vouga river basin (Portugal): poaching impact, preferential resting sites and spawning grounds. Hydrobiologia 582: 121–132.

Bell, A.M. 2005. Behavioral differences between individuals and two populations of stickleback (*Gasterosteus aculeatus*). J. Evol. Biol. 18: 464–473.

Bolnick, D.I., R. Svanback, J.A. Fordyce, L.H. Yang, J.M. Davis, C.D. Hulsey, and M.L. Forister. 2003. The ecology of individuals: incidence and implications of individual specialization. Am. Nat. 161: 1–28.

Broderson, J., P.A. Nilsson, L.A. Hansson, C. Ckov, and C. Bronmark. 2008. Condition-dependent individual decision-making determines cyprinid partial migration. Ecology 89: 1195–1200.

Caudill, C.C., W.R. Daigle, M.L. Keefer, C.T. Boggs, M.A. Jepson, B.J. Burke, R.W. Zabel, T.C. Bjornn, and C.A. Peery. 2007. Slow dam passage in adult Columbia River salmonids associated with unsuccessful migration: delayed negative effects of passage obstacles or condition-dependent mortality? Can. J. Fish. Aquat. Sci. 64: 979–995.

Claireaux, G. and C. Lefrancaois. 2007. Linking environmental variability and fish performance: integration through the concept of scope for activity. Philos. Trans. R. Soc. Lond., Ser. B: Biol. Sci. 362: 2031–2041.

Clemens, B.J., T.R. Binder, M.F. Docker, M.L. Moser, and S.A. Sower. 2010. Similarities, differences, and unknowns in biology and management of three parasitic lampreys of North America. Fisheries 35: 580–596.

Clemens, B.J., S. van der Wetering, J. Kaufman, R.A. Holt, and C.B. Schreck. 2009. Do summer temperatures trigger spring maturation in Pacific lamprey, *Entosphenus tridentatus*? Ecol. FW Fish 18: 418–426.

Congdon, P.D. 2010. Applied Bayesian Hierarchical Methods. CRC Press, 606 p.

Cote, D., L.M.N. Ollerhead, R.S. Gregory, D.A. Scruton, and R.S. McKinley. 2002. Activity patterns of juvenile Atlantic cod (*Gadus morhua*) in Buckley Cove, Newfoundland. Hydrobiologia 483: 121–127.

Docker, M. 2009. A review of the evolution of nonparasitism in lampreys and an update of the paired species concept. *In:* L.R. Brown, S.D. Chase, M.G. Mesa, R.J. Beamish, and P.B. Moyle [eds.]. Biology, management and conservation of lampreys in North America. Am. Fish. Soc. Symp. 72, Bethesda, Maryland pp. 71–114.

Fagan, W.F. 1997. Introducing a "boundary-flux" approach to quantifying insect diffusion rate. Ecology 78: 579–587.

Gelman, A., J.B. Carlin, H.S. Stern, and D.B. Rubin. 2003. Bayesian Data Analysis. Chapman & Hall/CRC.

Goodman, D.H., S.B. Reid, M.F. Docker, G.R. Haas, and A.P. Kinsiger. 2008. Mitochondrial DNA evidence for high levels of gene flow among populations of a widely distributed anadromous lamprey *Entosphenus tridentatus* (Petromyzontidae). J. Fish Biol. 72: 400–417.

Gross, M.R. 1991. Salmon breeding behavior and life history evolution in changing environments. Ecology. 72: 1180–1186.

Hess, J.E., N.R. Campbell, D.A. Close and M.F. Docker. 2012. Population genomics of Pacific lamprey: adaptive variation in a highly dispersive species. Mol. Ecol. (doi: 10.1111/mec.12150).

Jellyman, D.J., G.J. Glova, and J.R.E. Sykes. 2002. Movements and habitats of adult lamprey (*Geotria australis*) in two New Zealand waterways. N. Z. J. Mar. Freshw. Res. 36: 53–65.

Johnson, E.L., C.C. Caudill, M.L. Keefer, T.S. Clabough, C.A. Peery, M.A. Jepson, and M.L. Moser. 2012. Movement of radiotagged adult Pacific lampreys during a largescale fishway velocity experiment. Trans. Am. Fish. Soc. 141: 571–579.

Keefer, M.L., C.A. Peery, M.A. Jepson, and L.C. Stuehrenberg. 2004. Upstream migration rates of radiotagged adult Chinook salmon in riverine habitats of the Columbia River basin. J. Fish Biol. 65: 1126–1141.

Keefer, M.L., M.L. Moser, C.T. Boggs, W.R. Daigle, and C.A. Peery. 2009a. Variability in migration timing of adult Pacific lamprey (*Lampetra tridentata*) in the Columbia River, U.S.A. Environ. Biol. Fishes 85: 253–264.

Keefer, M.L., M.L. Moser, C.T. Boggs, W.R. Daigle, and C.A. Peery. 2009b. Effects of body size and river environment on the upstream migration of adult Pacific lampreys. N. Am. J. Fish. Manage. 29: 1214–1224.

Keefer, M.L., C.C. Caudill, C.A. Peery, and M.L. Moser. 2012. Context-dependent diel behavior of upstream-migrating anadromous fishes. Environ. Biol. Fishes (doi: 10.1007/s10641-012-0059-5).

Kelso, J.R.M and G.J. Glova. 1993. Distribution, upstream migration and habitat selection of maturing lampreys, *Geotria australis*, in Pigeon Bay Stream, New Zealand. Aust. J. Mar. Freshw. Mar. Res. 44: 749–759.

Kucheryavyi, A.V., K.A. Savvaitova, D.S. Pavlov, M.A. Gruzdeva, K.V. Kuzishchin, and J.A. Stanford. 2007. Variations of life history strategy of the Arctic lamprey *Lethenteron camtschaticum* from the Utkholok River (Western Kamchatka) J. Ichthyol. 47: 37–52.

Lin, B., Z. Zhang, Y. Wang, K.P. Currens, A. Spidle, Y. Yamazaki, and D. Close. 2008. Amplified fragment length polymorphism assessment of genetic diversity in Pacific lampreys. N. Am. J. Fish. Manage 28: 1182–1193.

Lucas, M.C., D.H. Bubb, M.H. Jang, K. Ha, and J.E.G. Masters. 2009. Availability of and access to critical habitats in regulated rivers: effects of lowhead barriers on threatened lampreys. FWBiol. 54: 621–634.

Mesa, M.G., J.M. Bayer, M.B. Bryan, and S.A. Sower. 2010. Annual sex steroid and other physiological profiles of Pacific lampreys (*Entosphenus tridentatus*). Comp. Biochem. Physiol., A: Mol. Integr. Physiol. 155: 56–63.

Miller, J.A., V.L. Butler, C.A. Simenstad, D.H. Backus, and A.J.R. Kent. 2011. Life history variation in upper Columbia River Chinook salmon (*Oncorhynchus tshawytscha*): a comparison using modern and ~500 year old archaeological otoliths. Can. J. Fish. Aquat. Sci. 68: 603–617.

Moser, M.L., P.R. Almeida, P. Kemp, and P.W. Sorenson. 2013. Spawning migration. Chapter 10. *In*: M. Docker [editor]. The Biology of Lampreys, SpringerVerlag.

Moser, M.L., A.L. Matter, L.C. Stuehrenberg, and T.C. Bjornn. 2002. Use of an extensive radio receiver network to document Pacific lamprey (*Lampetra tridentata*) entrance efficiency at fishways in the Lower Columbia River, USA. Hydrobiologia 483: 45–53.

Moser, M.L., D.A. Ogden, and B.P. Sandford. 2007. Effects of surgically-implanted transmitters in anguilliform fishes: lessons from lamprey. Journal of Fish Biology 71: 1847–1852.

Moser, M.L., R.W. Zabel, B.J. Burke, L.C. Stuehrenberg, and T.C. Bjornn. 2005. Factors affecting adult Pacific lamprey passage rates at hydropower dams: using "time to event" analysis of radiotelemetry data. *In*: M.T. Spedicato, G. Marmulla, and G. Lembo [eds.]. Aquatic Telemetry: Advances and Applications, FAOCOISPA, Rome pp. 61–70.

Noyes, C.J., C.C. Caudill, T.S. Clabough, D.C. Joosten, E.L. Johnson, M.L. Keefer, and G.P. Naughton. 2012. Adult Pacific lamprey migration behavior and escapement in the

Bonneville reservoir and lower Columbia River monitored using the juvenile salmonid acoustic telemetry system (JSATS), 2011. Technical Report 2012-4, Idaho Cooperative Fish and Wildlife Research Unit, University of Idaho, Moscow, Idaho.

Okubo, A. 1980. Diffusion and ecological problems: mathematical models. Springer-Verlag, New York.

Potter, I.C. 1980. The petromyzoniformes with particular reference to paired species. Can. J. Fish. Aquat. Sci. 37: 1595–1615.

Pulido, F. 2007. The genetics and evolution of avian migration. Bioscience 57: 165–174.

R Development Core Team. 2011. R: A language and environment for statistical computing. R Foundation for Statistical Computing, Vienna, Austria. ISBN 3-900051-07-0, URL http://www.R-project.org/.

Robinson,T.C., P.W. Sorensen, J.M. Bayer, and J.G. Seelye. 2009. Olfactory sensitivity of Pacific lampreys to a lamprey bile acid. Trans. Am. Fish. Soc. 138: 144–152.

Sih, A., A.M. Bell, J.C. Johnson, and R.E. Ziemba. 2004. Behavioral syndromes: An integrative overview. Q. Rev. Biol. 79(3): 241–277.

Spice, E.C., D.H. Goodman, S.B. Reid, and M.F. Docker. 2012. Neither philopatric nor panmictic: microsatellite and mtDNA evidence suggests lack of natal homing but limits to dispersal in Pacific lamprey.Mol. Ecol. 21: 2196–2930.

Stier, K. and B. Kynard. 1986. Movement of searun sea lampreys *Petromyxon marinus* during the spawning migration in the Connecticut River. Fish. Bull. U.S. 84: 749–753.

Su, Y. and M. Yajima. 2011. R2jags: A Package for Running jags from R. R package version 0.02-15. http://CRAN.R-project.org/package=R2jags.

Thorpe, J.E., M. Mangel, N.B. Metcalfe, and F.A. Huntingford. 1998. Modeling the proximate basis of salmonid lifehistory variation, with application to Atlantic salmon, *Salmo salar* L. Evol. Ecol. 12: 581–599.

Vrieze, L.A., R.A. Bergstedt, and P.W. Sorensen. 2011. Olfactory mediated stream finding behavior of migratory adult sea lamprey (*Petromyzon marinus*). Can. J. Fish. Aquat. Sci. 68: 523–533.

Waples, R.S., D.J. Teel, J.M. Myers, and A.R. Marshall. 2004. Life history divergence in Chinook salmon: historic contingency and parallel evolution. Evolution. 58: 386–403.

Yun, S., A.J. Wildbill, M.J. Siefkes, M.L. Moser, A.H. Dittman, S.C. Corbett, W. Li, and D.A. Close. 2011. Identification of putative migratory pheromones from Pacific lamprey. Can. J. Fish. Aquat. Sci. 68: 2194–2203.

Zabel, R.W. and J.J. Anderson. 1997. A model of travel time of migrating juvenile salmon, with an application to Snake River spring Chinook salmon. N. Am. J. Fish. Mgmt. 17: 93–100.

Zabel, R.W. 2002. Using "travel time" data to characterize the behavior of migrating animals. Am. Nat. 159: 372–387.

Zuur, A.F., E.N. Ieno, N.J. Walker, A.A. Saveliev, and G.M. Smith. 2009. Mixed effects models and extensions in ecology with R. Springer.

CHAPTER 7

Behavioral Ecology and Thermal Physiology of Immature Pacific Bluefin Tuna

Takashi Kitagawa

Introduction

There are eight tuna species in the genus *Thunnus*: albacore (*T. alalunga*), bigeye tuna (*T. obesus*), Atlantic bluefin tuna (*T. thynnus*), Pacific bluefin tuna (*T. orientalis*), southern bluefin tuna (*T. maccoyii*), yellowfin tuna (*T. albacares*), longtail tuna (*T. tonggol*), and blackfin tuna (*T. atlanticus*). Of these, albacore, bigeye tuna, Atlantic bluefin tuna, Pacific bluefin tuna, southern bluefin tuna, and yellowfin tuna are the most economically important species and are referred to as the principal market tunas because of their global economic importance and intensive international trade for canning and sashimi (Majkowski 2007).

In 2007, Japan's tuna catch ranked the highest in the world, accounting for 14% (248,000 tons) of the world's total tuna catch. In addition, Japan is the largest tuna consumer in the world and is supplied with 473,000 tons of tuna (the total of Japan's catch and imports) (Fishery Agency 2009). In particular, Pacific, Atlantic, and southern bluefin tuna contribute relatively less in terms of total catch weight of the principal market tunas; however, their individual value is high because they are used for sashimi, which is a raw fish delicacy in Japan and several other countries (Majkowski 2007).

Atmosphere & Ocean Research Institute, The University of Tokyo, Kashiwa, Chiba 277-8564, Japan.
Email: takashik@aori.u-tokyo.ac.jp

Still fresh in our memories is the fact that a Pacific bluefin tuna caught off northeastern Japan fetched a record 155.4 million yen or about $ 1.76 million in the first auction in January 2013 at Tokyo's Tsukiji fish market. The price for the 222 kg tuna beat last year's record of 56.49 million yen.

In contrast, bluefin tuna numbers have decreased by 80% or more since 1970 as a result of overfishing (Dalton 2005). In March 2010, at the 15th Conference of the Convention on International Trade in Endangered Species of Wild Fauna and Flora (CITES) in Doha, Qatar, a proposal to list Atlantic bluefin tuna in Appendix I of CITES proposed by Monaco was discussed and put to a vote. The proposed ban was voted down since fishery resources should be conserved and managed for sustainable use by Regional Fisheries Management Organization (RFMO), based on scientific stock assessments, rather than by CITES. Japan played a more leading role at the International Commission for the Conservation of Atlantic Tunas (ICCAT) and other RFMOs to prevent overfishing by adopting effective conservation and management measures based on scientific stock assessments, and by establishing a reliable monitoring system to ensure compliance by RFMO member countries.

Japan is the largest fishing nation as well as almost the exclusive consuming nation of Pacific bluefin tuna. Therefore, Japan has a special responsibility for sustainable use of this species. More than 70% of the total Pacific bluefin tuna catch is harvested by Japan, major Pacific bluefin tuna spawning grounds are near Japan, and most of the Pacific bluefin tuna harvested by other countries such as Mexico and Korea are exported to Japan (the Ministry of Agriculture, Forestry and Fisheries of Japan 2011: http://www.jfa.maff.go.jp/j/kokusai/kanri_kyouka/index2.html). Therefore, clarifying the movement and distribution of this species in detail is important for stock assessment and, if necessary, selection of regulatory measures. Recruitment estimation is critical for any tuna stock assessment in world oceans. The relationship between bluefin tuna abundance and various oceanic conditions has been reported (Uda 1957, 1973; Sund et al. 1981; Yamanaka 1982; Koido and Mizuno 1989; Matsumura 1989; Ogawa and Ishida 1989a, b; Bayliff 1994; Hamasaki and Nagai 1995; Itoh 2006, 2009; Yamada et al. 2006). This information includes the effect of water temperature on bluefin distributions and movement; however, most of the information is still very fragmented. Abundance of Pacific bluefin tuna stock is often estimated by age-structured population dynamics models such as virtual population analysis (VPA). However, VPA has lower reliability for young fish abundance estimates. In addition, the largest bluefin fisheries in the western Pacific target smaller fish, and it is estimated that up to 93% of the total fish landed are younger than three years of age (Itoh 2001).

Therefore, other reliable abundance indices for assessing current stock status must be identified (Yamada et al. 2006).

An acoustic (ultrasonic) tracking system has been used to directly measure Pacific bluefin tuna behavior (Marcinek et al. 2001). This system consists of a transmitter attached to the fish and a directional hydrophone and receiver system installed on a tracking vessel, allowing the depth at which the fish is swimming and the ambient water temperature to be transmitted to a receiver system on board. The location of the fish is also detected using a GPS on the vessel. Some of the limitations of studies using acoustic tags are as follows: (1) the duration of all successful tracks are less than one week. This is because of the practical difficulty in tracking a target continuously over long periods in rough seas, which often results in loss of the target; (2) only a single fish can be tracked at any one time; and (3) since fish may show abnormal behavior in the first few days after release, a longer monitoring period is essential to estimate their natural behavior, particularly for investigations of seasonal change and development of their behavior (Kitagawa et al. 2004a).

Since the end of the 1990s, electric tags such as "archival tags" have been developed and applied to tuna species. This recovery type tag enables time-series data to be recorded for a longer duration and at a higher resolution compared with previous acoustic tracking studies using various transmitters. The behavior of Pacific bluefin tuna has been observed using such tags, and thus, behavior, physiology, and ecology are being clarified gradually (Kitagawa et al. 2000, 2001, 2002, 2004b, 2006a, b, 2007a, b, 2009; Inagake et al. 2001; Block et al. 2003; Itoh et al. 2003a, b; Boustany et al. 2010; Boustany 2011) Marcinek et al. (Marcinek et al. 2001) and Farwell (Farwell 2001) reported on application of pop-up satellite archival tags to juvenile bluefin in the eastern Pacific. Domeier et al. analyzed data recovered from pop-up satellite archival tags, in addition to three surgically attached archival tags in juvenile bluefin tuna (Domeier et al. 2005). This type of tag detaches itself at a user preprogrammed time, floats to the surface, and transmits collected data via Argos.

In this chapter, the behavioral and thermal physiological characteristics of Pacific bluefin tuna are described. In particular, the effect of vertical temperature structure on their vertical distribution and movement, the influence of temporal and spatial changes in ambient water temperature on body temperature, their thermoconservation mechanism under low ambient temperatures, and their deep dives through the thermocline in relation to the occurrence of feeding events are discussed.

1. Horizontal Movements of Pacific Bluefin Tuna

1.1. Life history of the Pacific bluefin tuna

Sund et al., Bayliff, and Tsuji and Itoh reviewed the life history of the Pacific bluefin tuna based on results from fisheries data and conventional tagging experiments (Sund et al. 1981; Bayliff 1994; Tsuji and Itoh 1998).

The spawning grounds of the Pacific bluefin tuna are in the area between the Philippines and the Ryukyu Islands in the northwestern Pacific Ocean from April to June and in the Sea of Japan in August (Yabe et al. 1966; Ueyanagi 1969; Okiyama 1974; Kitagawa et al. 1995; Tanaka et al. 2006). Larvae hatched in the northwestern Pacific Ocean are carried by the Kuroshio Current, and juvenile bluefin tuna are transported near the coast of Japan. Age 0 fish, about 15 to 60 cm in fork length (FL), migrate north in summer and south in fall and winter along Japanese coastal waters (Yabe et al. 1966; Yukinawa and Yabuta 1967; Bayliff 1994). At age 0 or 1, some fish depart for a trans-Pacific migration, whereas others remain in the vicinity of Japan, migrating north and south seasonally (Tsuji and Itoh 1998).

Fish in the eastern Pacific also show seasonal migration, in which they are closer to the coast during summer and travel offshore in winter. The fish begin to move back to the western Pacific after staying in the eastern Pacific for a certain period of time, probably for spawning. The minimum observed size of fish at spawning is about 60 kg (about 150 cm in fork length (FL) at age 5, Shimose et al. 2009) along the Pacific coast of Japan and about 30 kg (age 3, Shimose et al. 2009) in the Sea of Japan. Batch fecundity ranges from 80,000 to 150,000 eggs/kg body weight (Tsuji and Itoh 1998).

Most adult fish remain in the western North Pacific after spawning. However, a small portion occasionally travel to the eastern North or South Pacific, off the coasts of Australia and New Zealand (Smith et al. 1994; Bayliff 1994), as small catches of Pacific bluefin tuna have been recorded in the South Pacific, particularly off Australia and New Zealand (Miyake et al. 2004). The following year, if they have not traveled too far, they presumably return to the spawning areas to spawn again (Bayiff 1994).

This information was inferred from fisheries data and conventional tagging experiments. However, more precise knowledge is needed on their migration and vertical and horizontal distributions.

1.2. Archival tagging

Archival tag logs include external and internal temperatures, swimming depth, and ambient light levels (Fig. 1). In addition, geo-locations have been estimated daily based on detection of the time of sunrise and sunset (Musyl et al. 2001; Ekstrom 2004). Because latitude estimations are often

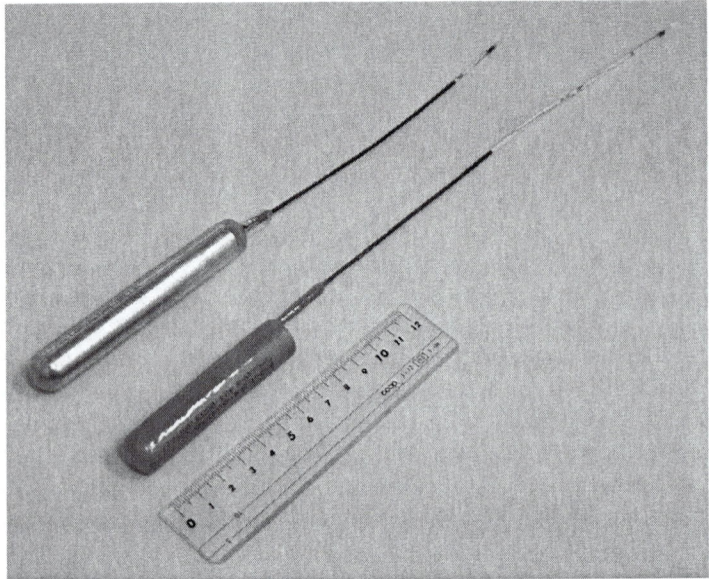

Fig. 1. Archival tags.

unreliable (Welch and Eveson 1999), latitude is usually adjusted using sea surface temperature (SST), as recorded in the summary file each day (Teo et al. 2004).

The microtechnology of archival tags is progressing rapidly. For example, the data memory of archival tags used by Kitagawa et al. (Kitagawa et al. 2000, 2001, 2002, 2004b, 2009) and Itoh et al. (Itoh et al. 2003a, b) was 256 KB; thus, external and internal temperatures, swimming depth, and ambient light levels were all measured at an interval of 128 s (675/d) for a maximum of 80 d, although daily records containing the date, estimated sunrise and sunset times, water temperatures at 0 m plus two other selectable depths (approximately 60 m and 120 m), and other information required to estimate location each day were recorded. In contrast, the memory of tags used by Kitagawa et al. was 8 MB (Kitagawa et al. 2007a), and the data memory of current archival tags is 128 MB. In addition, the sensors are more sophisticated, battery life has been extended, and the main body of the tag has become smaller and lighter. The latest tag specifications can be obtained from the websites of several electric tag companies (Lotek Wireless Inc., http://www.lotek.com/).

According to the results obtained from the archival tagging experiments in addition to fisheries data and conventional tagging experiments, Boustany briefly reviewed the life history of the Pacific bluefin tuna, including the Atlantic and southern bluefin tuna (Boustany 2011).

1.3. Horizontal distribution and movements of immature Pacific bluefin tuna in the western Pacific detected by archival tags

Pacific bluefin tuna with archival tags released off Tsushima in November or December remain in the East China Sea (Kitagawa et al. 2000, 2001, 2002, 2004b, 2009; Inagake et al. 2001; Itoh et al. 2003a, b), which is a well-known wintering habitat for immature bluefin tuna.

Three typical tracks are shown in Fig. 2. Bluefin 177 which remained within the East China Sea (Fig. 2a), as did most of the other individuals reported in studies by Kitagawa et al. (Kitagawa et al. 2000, 2004b) and Itoh et al. (Itoh et al. 2003a, b), although the differences in timing of horizontal

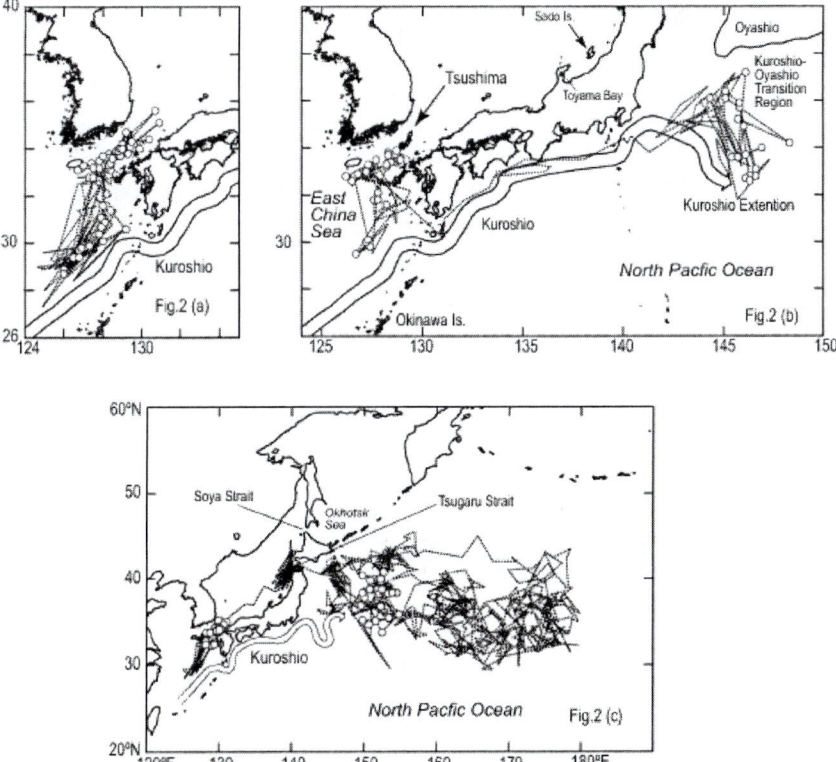

Fig. 2. Estimated archival tag tracks for a Pacific bluefin tuna swimming in the East China Sea (Bluefin 177) (a) and for bluefin migrating into the Kuroshio-Oyashio transition region [(b) Bluefin 209 and (c) Bluefin 199]. Open circles indicate positions where time-series data were recorded. Solid circle indicates the release site of Pacific bluefin tuna with archival tags (modified from Kitagawa et al. 2004b).

movement varied among years because of changes in SST. The high-density distribution area shifted to the south of Tsushima Islands during winter, although some fish remained off the Tsushima Island; more fish moved to the Kuroshio frontal area in the East China Sea. In April, when the Tsushima Warm Current and the Kuroshio Current were warmer, which brought warmer waters to the north, fish distribution also shifted to the north (Kitagawa et al. 2006b). Similar to Bluefin 177, most of the tagged fish remained within the East China Sea and were recaptured close to Tsushima Island from May to June, when they moved up from the south.

In contrast, Bluefin 209 (Fig. 2b) migrated further into the western North Pacific and the Sea of Japan. The fish migrated into the western North Pacific on 7 March, 1996 after moving within the East China Sea for about three months. They then moved eastward straight along the coastal side of the Kuroshio front and migrated into the Kuroshio-Oyashio transition region, which is characterized by irregularly distributed eddies and thermohaline fronts between the Kuroshio extension and the subarctic Oyashio front (Kawai 1972) in April 1996. This fish was found in the southeast and northwest sections of this area from April to June and was recaptured by purse seine on 7 June, 1996. Bluefin 164 entered the Sea of Japan and changed phases, moved horizontally and linearly toward the northeast in the Sea of Japan, and was recaptured in Toyama Bay (Kitagawa et al. 2002).

Some tagged fish entered the Sea of Japan, where the oceanographic characteristics are quite different from those of the East China Sea. A few fish were recovered in the Sea of Japan (by set nets in Toyama Bay or Sado Island), but a few fish such as Bluefin 199, shown in Fig. 2c, moved into the Kuroshio-Oyashio transition region in the western North Pacific. The fish migrated through the Sea of Japan and the Tsugaru Strait in November 1996, after staying in the East China Sea for six months (Fig. 2c). The fish stayed in the Kuroshio-Oyashio transition region for about two months just before recapture. One of the tagged fish that entered the Kuroshio-Oyashio transition region went through the Soya Strait in August and the Okhotsk Sea before entering the region (Inagake et al. 2001).

Detailed analysis of their horizontal movements in the area, in relation to oceanographic conditions such as SST and chlorophyll-a, was reported by Inagake et al. (Inagake et al. 2001). The fish recaptured in the area had grown to FL of 144–150 cm at recovery (Inagake et al. 2001; Kitagawa et al. 2004b). These fish preferred warm water of about 18°C in the upper layer of this area. They showed clockwise migration patterns closely related with the ocean structure in and around the region. The fish moved westward in spring, in and around the Kuroshio extension, northward in summer in the warmer water, and spread from the crest of the Kuroshio extension, eastward in fall along the south of the Oyashio front, and southward in early winter to the Kuroshio extension. It is likely that their migration routes in this area are related to changes in chlorophyll-a concentration and ocean currents (Inagake et al. 2001).

1.4. Trans-Pacific migration from the western to eastern North Pacific

The other fish traveled from the western Pacific Ocean to the eastern Pacific Ocean as follows (Fig. 3; Itoh 2003a; Kitagawa 2009). The fish was released off Tsushima on 29 November, 1996 at 55 cm FL and remained for a period within the East China Sea. It moved to the Pacific Ocean on 1 May, 1997 and then traveled eastward straight from a position off the south coast of Kyushu and migrated into the Kuroshio-Oyashio transition region in the western Pacific in May 1997 (Itoh et al. 2003a). Although the fish stayed in the region for some time and shifted habitat areas to the Oyashio frontal area, it initiated its trans-Pacific migration on 11 November, 1997, moving eastward (Mean direction: 3.8° south of east) along the Subarctic Frontal Zone (V test for circular uniformity, $P < 0.0005$; Kitagawa et al. 2009). The fish changed its direction of movement to the southeast on 8 December, 1997. It arrived in the eastern Pacific on 15 January, 1998 and was recaptured 610 days after release by a recreational fishing vessel on 1 August, 1998 off Baja California, Mexico, at FL of 87.6 cm.

As indicated by Itoh et al. (Itoh et al. 2003a), migration of immature Pacific bluefin tuna appears to consist of a residency phase, comprising more than 80% of all days as a randomly distributed phase, and a traveling or directional movement phase. However, the mechanisms for the phase change (sit-and-go and/or go-and-sit) have not been clarified, although Polovina proposed the hypothesis that migration of juvenile bluefin into the eastern Pacific increases in years when the abundance of sardines off Japan declines (Polovina 1996).

1.5. Movements of immature Pacific bluefin tuna in the eastern North Pacific

Kitagawa et al. (Kitagawa et al. 2007a) and Boustany et al. (Boustany et al. 2010) revealed the seasonal movements of juvenile bluefin tuna off the west coast of North America (in the eastern North Pacific). Electronically tagged juvenile Pacific bluefin tuna were released off Baja California in the summer of 2002 (Kitagawa et al. 2007a) and off the coast of California, USA, and Baja California, Mexico, between August 2002 and August 2005 (Boustany et al. 2010).

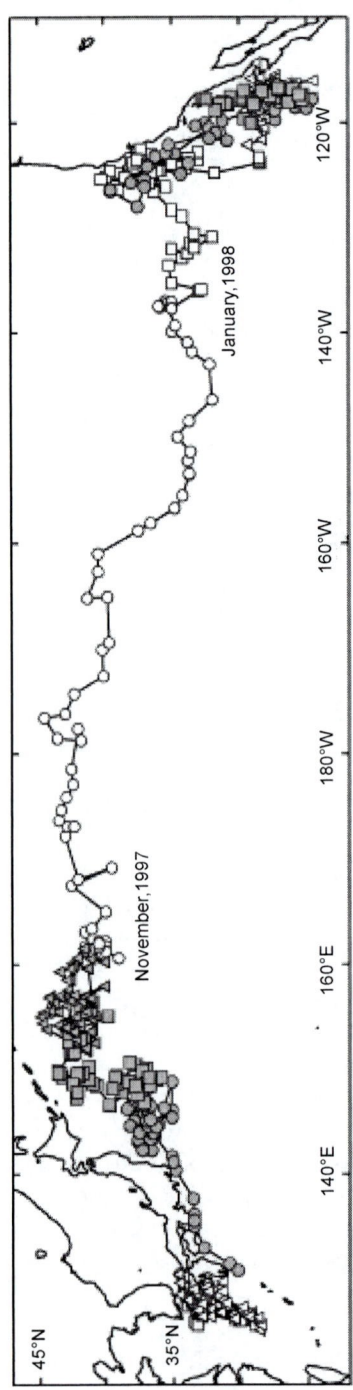

Fig. 3. Estimated archival tag tracks for a Pacific bluefin tuna that made a trans-Pacific migration (Bluefin 241). It should be noted that symbols are changed every 2 mon (modified from Kitagawa et al. 2009).

Electronically tagged bluefin tuna showed repeatable seasonal movements along the west coast of North America (Fig. 4). The fish showed latitudinal movement patterns that were correlated with peaks in coastal upwelling-induced primary productivity and Pacific sardine (*Sardinops sagax*) availability (Domeier et al. 2005; Kitagawa et al. 2007a; Boustany et al. 2010).

In the spring through fall, bluefin tuna were located in areas with the highest levels of primary productivity available in the California Current ecosystem. In summer, the fish were located primarily in the Southern California Bight and over the continental shelf of Baja California. In autumn, the landings of Pacific sardines increased to their highest level in central California coincident with a rising SST (Goericke et al. 2004). This period was the beginning of the warmest SSTs in the annual cycle of the region and suggests why the Pacific bluefin moved up the coast. The northward

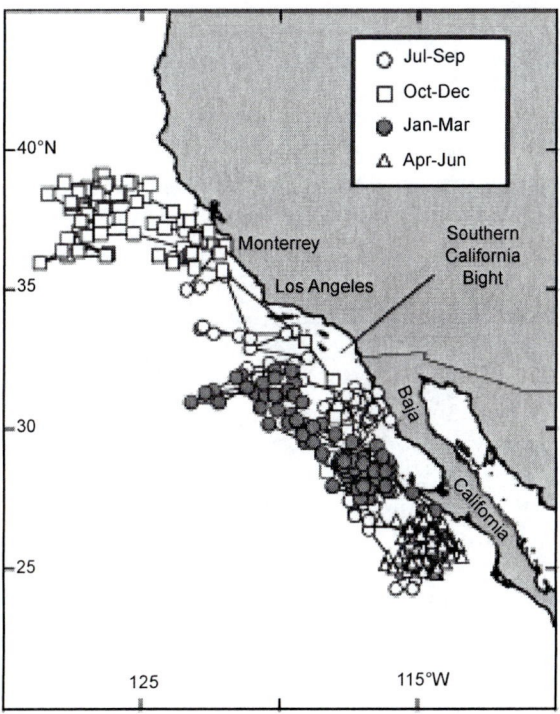

Fig. 4. Estimated track of Bluefin 315 from daily geo-location positions (modified from Kitagawa et al. 2007a).

movement of prey because of high productivity and warmer temperatures is accompanied by a weakened equatorward wind stress in the region, leading to a northward migration of Pacific bluefin (Kitagawa et al. 2007a).

In mid-winter, the tagged bluefin tuna were found in areas with lower productivity compared with other regions along the coast at that time of year, suggesting that during the winter, bluefin tuna feed on aggregations of pelagic red crabs, sardines, and anchovies that preferentially spawn in areas of reduced coastal upwelling (Boustany et al. 2010).

1.7. Trans-Pacific migration from the eastern to western Pacific

According to Boustany et al. (Boustany et al. 2010), of the archival tagged fish released between August 2002 and August 2005, 17 moved offshore, which was defined as greater than 5° from the North American coast. Seven (4.5%) fish moved across the Pacific and crossed the dateline at 180°W longitude. Two trans-Pacific fish subsequently moved back to the eastern Pacific, and one individual traveled back to the western Pacific where it was recaptured. All other tagged fish remained in the North American coastal region, within 5° of shore, and the vast majority of daily geo-locations were in this region. Offshore movements (>5° from shore) by individual fish started in the late fall through winter in all years, and the number of fish traveling offshore in the winter months varied among years with eight in 2003, six in 2004, and three in 2005.

Archival tagging studies on the trans-Pacific migration from the eastern to western Pacific (Block et al. 2003; Boustany et al. 2010) revealed that Pacific bluefin tuna make use of more southern waters compared with migration from the western to eastern Pacific (Block et al. 2003; Itoh et al. 2003a; Kitagawa et al. 2009). During the trans-Pacific migration, as shown by Block et al. (Block et al. 2003), they remain at seamount areas for some time, when they are in the residency phase defined by Itoh et al. (Itoh et al. 2003a). Immature fish in the eastern Pacific return to the western Pacific, although only mature fish return to the western Pacific (Bayliff 1994). The importance and evolutionary process of the trans-Pacific migration, which covers more than 8,000 km between the western and eastern Pacific, has not been clarified.

2. Vertical Movement and Distribution of Immature Pacific Bluefin Tuna

Pacific bluefin tuna not only move horizontally but also vertically to seek prey and avoid predators. Therefore, in their immature stages, vertical movement is crucial for the purpose of archiving their migration during the Pacific life cycle. The difference in their vertical movement according to residency area and their vertical distribution and movements are described in this section. In particular, the effect of vertical temperature structure on vertical distribution and movement, the influence of temporal and spatial changes in ambient water temperature on body temperature, their thermoconservation mechanisms under low ambient temperatures, and their deep dives through the thermocline in relation to the occurrence of feeding events are discussed.

2.1. Vertical movements in the East China Sea

Vertical movements: Time-series data obtained from Bluefin 177, which swam in coastal waters of Tsushima in the East China Sea in winter (15 December, 1995 to 21 January, 1996) and in summer (24 May, 1996 to 30 June, 1996), are shown in Fig. 5 (Kitagawa et al. 2000). Although the bluefin made diel vertical migrations up to 120 m depth during the daytime in winter (November–January) and stayed at shallower depths at night, the ambient temperature did not change much in association with the large vertical migration (Fig. 4a). This small change is attributed to strong vertical mixing due to surface cooling. The right panel of Fig. 5a shows the frequency distributions of the swimming depth and the mean vertical distribution of ambient temperature during the same period, recorded by the tag attached to Bluefin 177. In December, vertical profiles of ambient temperature (about 17.4°C) seemed almost homogeneous above a 100-m depth.

In contrast, summer (March–June) was characterized as a stratified season (Fig. 5b). The bluefin tuna spent most of their time at the surface during the daytime and at night, suggesting that a seasonal thermocline possibly regulates vertical distribution of the bluefin tuna, but they often repeated brief dives below the thermocline during the daytime; the summer maximum diving depth reached about 120 m, the same as that in winter. However, average daytime depth (less than 1 m, Fig. 5b) was shallow and not much different from that at night (3.9 m) (Kitagawa et al. 2000). Diving resulted in a marked change in ambient temperature because of the existence of a seasonal thermocline below the surface mixed layer.

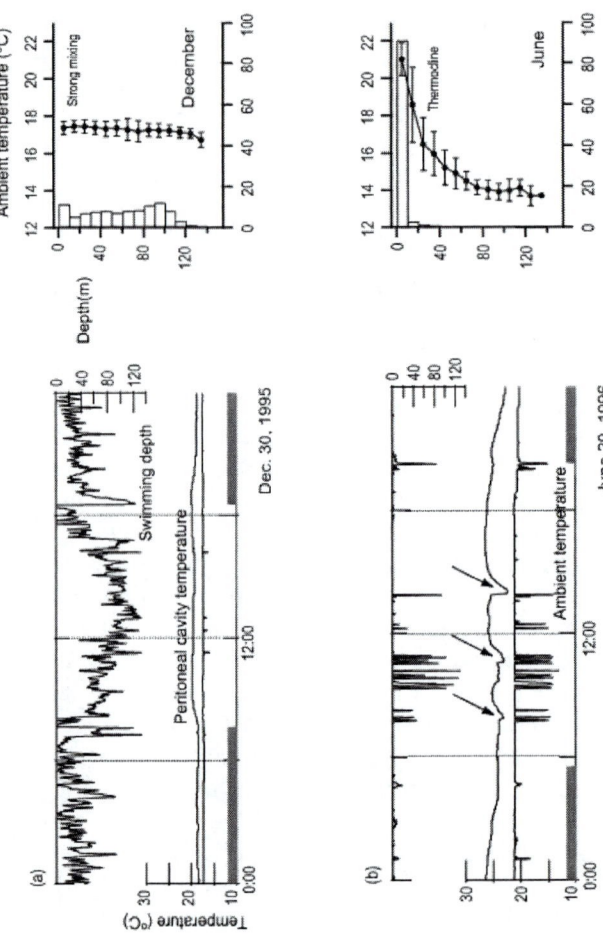

Fig. 5. (a) Time-series depth (upper), peritoneal cavity temperature (middle), and ambient temperature (lower) data on 30 December 1995, obtained from Bluefin 177 swimming in the East China Sea. Shaded time zones indicate night time (left panel). Frequency distributions of the swimming depth and vertical profiles of mean ambient temperature (and standard deviation, SD) during the daytime are shown in the right panel, modified from Kitagawa et al. (2000). (b) Time-series depth (upper), peritoneal cavity temperature (middle), and ambient temperature (lower) data on 20 June 1996, obtained from Bluefin 177. Shaded time zones indicate night time (left panel). Arrows indicate temperature drops in the peritoneal cavity due to feeding. Frequency distributions of swimming depth and vertical profiles of mean ambient temperature (SD) during the daytime are shown in the right panel (modified from Kitagawa et al. 2000, 2004a).

2.2. Relationship among swimming depth, dive duration, and ambient temperature

Figure 6 shows the relationship between the thermal gradient and frequency of bluefin swimming in the surface waters (0 m–10 m) during the daytime (Kitagawa et al. 2000, 2002). The thermal gradient was defined as the difference between the average temperature at the surface and that at depths of 50 m–60 m, indicating the relative strength of the thermocline. When the thermal gradient became steeper, the bluefin spent a much longer time at the surface, suggesting that the vertical ambient temperature structure apparently affects bluefin swimming depth (Fig. 5). In fact, however, the average surface temperature was 16.3°C–21.5°C, which was significantly correlated with the frequency of bluefin swimming at the surface water during the daytime. This suggests that there are interactions between

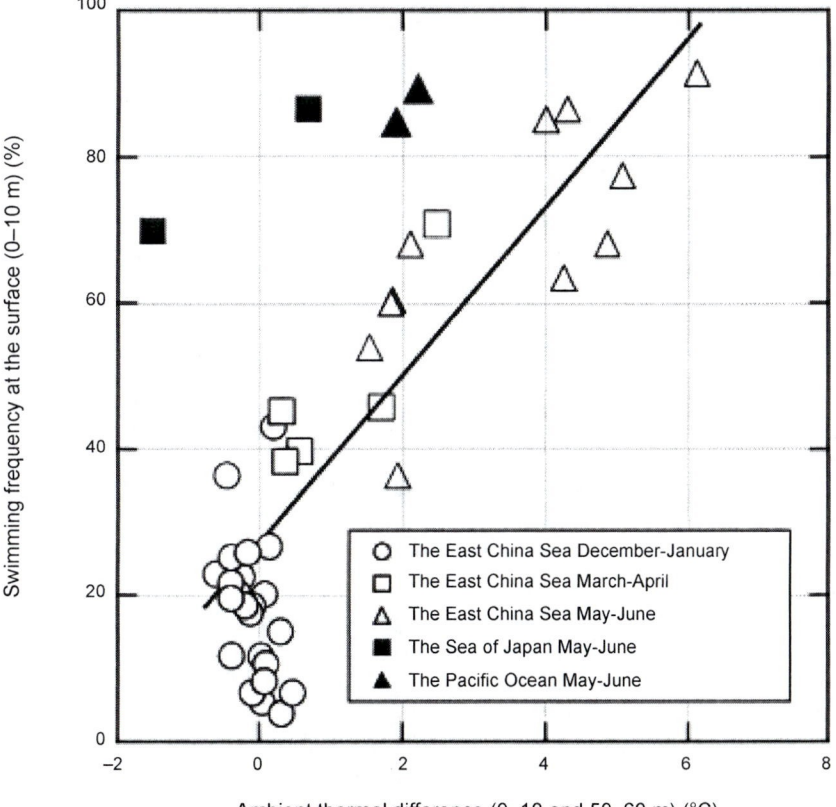

Fig. 6. Relationship between the thermal difference (0–10 and 50–60 m) and the frequency of swimming at the surface (0–10 m; modified from Kitagawa et al. 2002).

swimming frequency at the surface, the thermal gradient, and the average SST (Kitagawa et al. 2000). Thus, partial correlations among these three factors were examined, and the correlation between each pair of the three factors was calculated while the values of the other factor were held constant (Zar 1996). As a result, the correlation shown in Fig. 6 was significant (partial correlation coefficient = 0.839, $P < 0.001$, Kitagawa et al. 2000), whereas the correlation between the average surface water temperature and the surface swimming frequency was not significant (partial correlation coefficient = -0.085, $P > 0.1$, Kitagawa et al. 2000).

Kitagawa et al. further showed a relationship between the mean thermal gradient and mean dive duration together with the mean dive frequency (Kitagawa et al. 2004, 2007b). Correlations between each pair of the four factors (dive duration, dive frequency, thermal gradient, and average SST) were examined. All dive durations during winter were for more than 1000 s (Kitagawa et al. 2004), whereas dive durations became shorter in the period March–June, when the thermal gradient became steeper. In particular, dive duration in June decreased to less than 10 min, except for one point in June. These results strongly suggest that bluefin tuna tend to avoid vertical thermal changes at the thermocline, with the exception of frequent dives.

However, it should be noted that although dive frequency was more than 10 times per day with intervals of 1.2 hr on an average during March–May, it decreased in June when the gradient became steeper (Kitagawa et al. 2007b; Fig. 6). In addition, the power spectra of vertical moments for all fish were high for less than 3.0 hr in March–May, indicating that dive frequency (10 times/d) could be transformed into a 1.2 hr cycle of movements in the power spectra, but the spectra decreased in June. Therefore, dive frequency may not be related to thermal gradient strength.

2.3. The Kuroshio-Oyashio Transition Region

Pacific bluefin tuna swimming during the summer in the Kuroshio-Oyashio transition region mostly remained at the surface during the daytime and at night and rarely made dives to depths of more than 100 m during the daytime (Figs. 6 and 7, Kitagawa et al. 2002, 2004b). The frequency for all individuals at the surface was 66.8%–94.0%, which was significantly greater than that in the East China Sea (Kitagawa et al. 2004). These observations suggest that larger bluefin feed at the surface. Small fluctuations in the surface ambient water temperature were observed for a few bluefin, which was attributed to their swimming in a homogeneous water mass in the Kuroshio-Oyashio transition region. In fact, one of the fish was mostly located on the Kuroshio front and in an eddy generated from the Kuroshio extension in the Kuroshio-Oyashio transition region (Kitagawa et al. 2002, 2004b). The information available for this region is lesser than

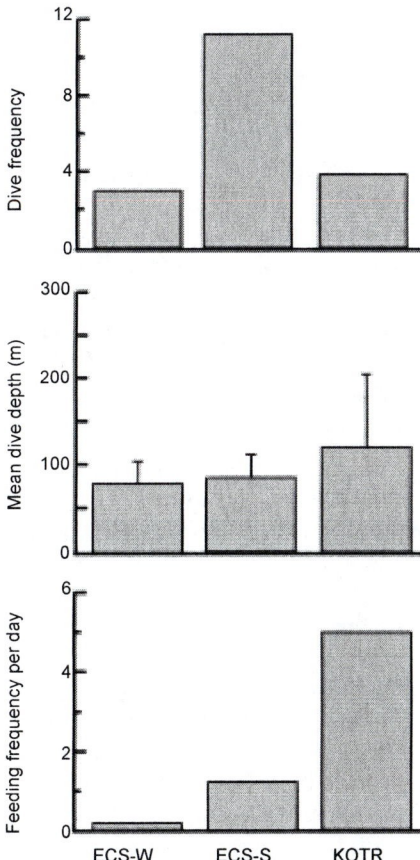

Fig. 7. Left columns: dive frequency, mean dive depth, and feeding events during the daytime for Bluefin 177 during December–January in the East China Sea (ECS-W). Middle columns: same except for Bluefin 177 during May–June in the East China Sea (ECS-S). Right: same as Bluefin 177, except for Bluefin 209 during May–June in the Kuroshio-Oyashio transition region (KOTR) (modified from Kitagawa et al. 2004b).

that available for the bluefin's vertical movements in the East China Sea. Thus, it is important to obtain more data in this region.

2.4. Sea of Japan

Some tagged fish entered the Sea of Japan, which has oceanographic characteristics quite different from those of the East China Sea. More than 90% of the Sea of Japan has temperatures exceeding 5°C (Yasui et al. 1967), and prey fauna in this sea are scarce compared with those in the Kuroshio region in the western North Pacific (Nishimura 1965). The fish made frequent dives during the daytime but few dives at night. The frequency of dives more than 100 m was less than that for fish in the East China Sea (Kitagawa et al. 2002; Fig. 6). As the fish entered the Sea of Japan horizontally, they moved linearly toward the northeast , and the ambient SST decreased slightly to 11°C (Kitagawa et al. 2002), which was much lower than that for fish such as Bluefin 177 in the East China Sea (e.g., Fig. 5). The fish dives made in May showed that the ambient temperature was often less than 5°C. However, the ambient temperature did not change markedly thereafter, even during large vertical migrations. This suggests that the fish migrated from the marginal region of the East China Sea, where there was strong vertical thermal stratification, to the Sea of Japan, where strong vertical mixing due to winter cooling prevailed. Information on bluefin vertical movements in the Sea of Japan is less than that available for the East China Sea, and thus, it will be important to obtain more data in this region.

3. Thermoconservation Mechanisms Inferred from Body Temperature in Free-swimming Pacific Bluefin Tuna

Thunnus is a well-known fish genus with an elevated body temperature (Kishinoue 1923; Barrett and Hester 1964; Carey et al. 1971, 1984; Carey 1973; Stevens et al. 1974, 2000). Therefore, it is of great importance to clarify the mechanisms that regulate body temperature under ambient water temperatures, which can change considerably as bluefin move horizontally and vertically. Hence, the thermoconservation mechanisms of free-swimming immature bluefin under low ambient temperatures are described based on the results of Kitagawa et al. who analyzed ambient water (T_a) and body temperature (T_b) records obtained from archival tags (Kitagawa et al. 2001, 2006a, 2007b).

3.1. Relationship between ambient water and body temperature

The relationship between the mean ambient water temperature and body (peritoneal cavity) temperature for 640 s (= sampling interval: 128 s × 5) was examined by Kitagawa et al. based on the time range of dive duration during a stratified season (May–June) (Kitagawa et al. 2001). The range in ambient water temperature was wider in June, because of repeated dives through the thermocline during the daytime. The temperature differences were much larger than those in December. The differences increased as ambient water temperatures decreased, and significant negative correlations were found for all individuals. This indicates that body temperature is maintained on a short time scale (640 s) when the bluefin makes repeated dives through the thermocline into depths with lower ambient temperature (Kitagawa et al. 2001).

3.2. Heat budget model

A heat budget model was used to examine the varying relationship between ambient water and peritoneal cavity temperature. Heat loss or gain (i.e., storage of heat) is proportional to the difference between body cavity and ambient water temperatures (Holland et al. 1992; Brill 1996; Brill et al. 1994; Kitagawa et al. 2001; Kitagawa et al. 2006a; Kitagawa et al. 2007b), as demonstrated in the following equation:

$$
\left.\begin{aligned}
\frac{dT_b}{dt} &= \lambda(T_a - T_b) + \dot{T}_m \\
\lambda &= k + w
\end{aligned}\right\} \text{, Equation (1)}
$$

where λ is the whole-body heat-transfer coefficient (/s, note that in some previous studies, k was used as the whole-body heat-transfer coefficient), w is arterial blood flow (/s, Kitagawa and Kimura 2006), $\dot{T}_m$ is the rate of temperature change due to internal heat production (°C/s), T_a is ambient water temperature (°C), and T_b is body (peritoneal cavity) temperature (°C). The first term on the right-hand side corresponds to conductive heat exchange. This equation indicates that heat production and conductive heat exchange play important roles in body temperature fluctuations in bluefin.

The λ and $\dot{T}_m$ values for all the fish have been estimated (Kitagawa et al. 2002, 2006a) from data fluctuations between ambient and body (peritoneal cavity) temperatures obtained from archival tags. In addition, Kitagawa et al. calculated the heat-transfer time ($1/\lambda$), which is the inverse value of λ, and is the time required for T_b to reach $1/e$ ($e = 2.71828…$: Napier's constant) (63%) of the new T_b value (Kitagawa et al. 2007b).

3.3. Thermoconservation mechanisms of immature Pacific bluefin tuna

According to Holland et al. bigeye tuna undergo physiological and behavioral thermoregulation by changing the heat-transfer coefficient λ by two orders of magnitude ($\lambda = 5.22 \times 10^{-4}$ for cooling, 4.01×10^{-2} for warming) (Holland et al. 1992). Disengaged heat exchangers allow rapid warming as the fish ascends from cold water into warm surface water, and they are reactivated to conserve heat when the fish returns to cold water. It is evident from their analysis that the heat exchangers of bigeye tuna are disengaged to allow rapid warming as they ascend from cold water into warmer surface waters and are reactivated to conserve heat when they return to the depths.

However, there is almost no difference in the λ values for immature Pacific bluefin tuna, (Bluefin 177: $\lambda = 3.7 \times 10^{-4}$ for cooling, 3.6×10^{-4} for warming, Kitagawa et al. 2001), suggesting that bluefin do not undergo physiological thermoregulation like bigeye. Rather, internal heat production $(\dot{T}_m)$ contributes much more than heat transfer to body temperature fluctuations in immature bluefin tuna. All $\dot{T}_m$ values (1.09×10^{-3} to 3.12×10^{-3}°C/s) estimated by Kitagawa et al. (Kitagawa et al. 2001) were higher than those of bigeye tuna described by Holland et al. (Holland et al. 1992). Since this higher heat production makes the dT_b/dt term in Eq. (1) larger when both λ and T_a are constant, it could relieve the reduction in body temperature due to low ambient water temperature. It should be noted that λ for mature Atlantic bluefin tuna increased rapidly at the high ambient temperatures encountered in the Gulf of Mexico and was significantly higher at night in the breeding phase, when the fish were exhibiting shallow oscillatory dives, suggesting that they were behaviorally and physiologically thermoregulating in the Gulf of Mexico (Teo et al. 2007).

In contrast, the heat-transfer time $(1/\lambda)$ for all fish was more than 40 min, except for one case (Kitagawa et al. 2007b), indicating that it takes a long time to raise T_b, which decreases by a dive to the same level as before the dive, if they dive to depths through the thermocline for a long period, and T_b would drop and be close to the level of T_a. To summarize, the bluefin tuna may thermoregulate behaviorally by spending most of their time at the surface during the stratified season in the East China Sea; also, repeated brief dives to depths through the thermocline prevent the bluefin from T_b heat loss (Kitagawa et al. 2007b).

4. Dives Through the Thermocline in Relation to the Occurrence of Feeding Events

Block et al. interpreted diving below the thermocline by yellowfin tuna as a behavior to avoid predators or ships (Block et al. 1997). In contrast, Holland et al. (1992) considered bigeye tuna diving to be for foraging purposes (Holland et al. 1992). The purpose of diving in immature bluefin is perhaps the latter because the frequency of diving is quite high and occurs mostly during the daytime.

Occurrence of feeding events inferred from thermal fluctuations in the peritoneal cavity recorded by archival tags has been investigated by Kitagawa et al. (Kitagawa et al. 2004b, 2007a, b). Figure 4b shows part of the time-series data of ambient water and peritoneal cavity temperatures as well as swimming depths for Bluefin 177. In this figure, the arrows indicate decreases in peritoneal cavity temperature. Ingestion of food, which results in a decline in temperature either from the prey or from water entering the peritoneal cavity during a feeding event, provides a foraging event marker (Carey et al. 1984; Gunn and Block 2001; Itoh et al. 2003b; Kitagawa et al. 2004b, 2007a, b). It should be noted that gradual increase in temperature after decline in temperature refers to specific dynamic action (the heat increment of feeding), defined as the energy expended (or heat produced) during ingestion, digestion, absorption, and assimilation of a meal. It is likely to elevate metabolism for many hours following a meal and is therefore likely to represent a significant component of the daily energy budget of bluefin tuna (Carey et al. 1984; Korsmeyer and Dewar 2001; Bestley et al. 2008; Clark et al. 2008, 2010).

Kitagawa et al. considered the decreases in peritoneal cavity temperature to be an indicator of a feeding event only when the temperature change could not be explained by changes in ambient water temperature (Kitagawa et al. 2004b, 2007a, b). Kitagawa et al. defined feeding depth as the depth at which a feeding event occurs (Kitagawa et al. 2007a, b).

4.1. The East China Sea

Mean monthly feeding events increased from March–June in the East China Sea (Kitagawa et al. 2004b, 2007b). Feeding depths were also less than 50 m in June (Kitagawa et al. 2007b). These observations indicate that food availability was greater in June than in other months. Considering dive frequency, it could be assumed that food biomass availability at the surface is greater in June than in March–May and that bluefin obtain food at the surface in June through horizontal movements (Kitagawa et al. 2007b). As Kitagawa et al. reported previously, vertical diving activity could be related to physical conditions such as light intensity (Kitagawa et al. 2001). For example, Bluefin 177 made fewer dives during the daytime, when solar radiation was comparatively low. However, vertical movements for almost all fish decreased in June, irrespective of changes in solar radiation (Kitagawa et al. 2007b). Hence, the effect of light intensity could reinforce the decrease in vertical movement or it would be unnecessary for bluefin to make vertical movements because of food availability at the surface. Conversely, bluefin repeatedly dive to depths through the thermocline in March–May because adequate food biomass is unavailable at the surface. Bluefin tuna with sizes of 20 cm–65 cm consume anchovies (Yamanaka et al. 1963). Although the anchovy is a pelagic fish, its swimming depth often extends deeper in the daytime during spring/summer. The greatest recorded swimming depth for anchovies is 125 m in spring and 70 m in summer (Nozu 1966). Other pelagic species such as sardines, common squid, and round herring are often distributed at some depth (Yokota 1961;

Hamabe 1964; Suzuki et al. 1974; Aoki and Murayama 1993). *Engraulis japonicus* is a pelagic fish that aggregates at the surface in the East China Sea in May–June (Ohshimo 1996). Anchovies may also be restricted to the surface in June when the thermocline becomes steep. It is noteworthy that the demersal lightfish *Maurolicus muelleri* has been found in bluefin stomachs (Dragovich 1969). This species is dominant in the micronekton of the East China Sea and is distributed below the thermocline during the daytime (Ohshimo 1998). This finding could be further evidence that bluefin dive through the thermocline to find food in March–May.

To summarize, immature Pacific bluefin tuna in the East China Sea have two contradictory demands in March–May when adequate food biomass is unavailable at the surface: (i) they want to feed in colder depths below the thermocline and (ii) they want to maintain a high body temperature, perhaps for rapid growth (Brill 1996). They simultaneously fulfill both requirements by undertaking repeated dives for short periods. Thus, such dives are an adaptive behavior to seasonal changes in the vertical thermal structure of the water column. The fish compensate for the need to feed in colder depths, while maintaining a high body temperature, by undertaking multiple short duration dives rather than fewer longer duration dives.

4.2. The Kuroshio-Oyashio transition region and the Eastern North Pacific

Feeding events during summer in the Kuroshio-Oyashio transition region are much more frequent than those in the East China Sea (Kitagawa et al. 2004b), despite the lower diving frequency. The mean horizontal distance traveled was also significantly higher, and it seemed that the fish may move horizontally to feed on prey accumulating at the surface more in this area than in the East China Sea, which is probably one of the reasons why they migrate here from the latter region.

One bluefin tuna migrated to the Oyashio frontal area, where both the horizontal and vertical thermal gradients are much steeper. The fish spent most of the time on the warmer side of the front and often traveled horizontally to the colder side during the daytime, perhaps to feed. This suggests a thermal barrier effect from the Oyashio front (Kitagawa et al. 2004b).

A number of anchovies have been caught at depths shallower than 25 m in the northern Kuroshio extension region (Takahashi et al. 2001). The Kuroshio extension also has abundant prey such as such sardines, mackerel, and lantern fish (Kinoshita 1998) as well as pelagic crustaceans such as krill of the genus *Euphausia* (Endo 2000), which are bluefin food (Kishinoue 1923). These reports suggest that the vertical distribution of prey for bluefin may be significantly different in the East China Sea and the

Kuroshio-Oyashio transition region and that this may lead to the difference in bluefin swimming depths between these two regions. Anchovy biomass at the surface is much larger in the transition region than in the East China Sea. For example, the average total annual catch of Japanese anchovy in this region in 1995–1999 was 100,527 tons, whereas that in the East China Sea was 47,219 tons (The Ministry of Agriculture, Forestry and Fisheries, 1997–2001).

Fish are primarily located in the Southern California Bight and over the continental shelf of Baja California during summer (Kitagawa et al. 2007a; Boustany et al. 2010), where juvenile Pacific bluefin use the top of the water column and undertake occasional, brief forays to depths below the thermocline. Bluefin migrate north to the waters off the central California coast in autumn, when thermal fronts form as the result of weakened equatorward wind stress. Examination of ambient and peritoneal temperatures revealed that bluefin tuna probably feed on Pacific sardines along the frontal boundaries (*S. sagax*) during this period. The bluefin return to the Southern California Bight during mid-winter, possibly because of strong downwelling and depletion of prey species off the central California waters.

Conclusion

Fish behavior, such as vertical movements and (horizontal) migration, is considered a response accompanied by change in physiology to a stimulus from the environment. Therefore, evaluating data that simultaneously measure fish behavior together with their environments will lead to an understanding of the interaction between their physiological state, behavior, and environments.

From the data recorded by archival tags and the ambient temperature structure, we clarified that the vertical and horizontal distribution of prey species plays an important role in the feeding behavior of Pacific bluefin tuna. Archival tags (as bio-logging science) provide fisheries-independent measurement of fish behavior including environmental and physiological information. Bio-logging is also a useful technique to collect various types of environmental information, and it provides information even with the time and space in which observations by research vessels are impossible. Therefore, if tags can be retrieved, it will be possible to collect detailed information for longer durations and at higher resolutions compared with previous fisheries data analysis or acoustic tracking studies. As a responsible fishery and consumption country, new knowledge about the ecology of Pacific bluefin tuna brought about by this technology serves as a biological basis in Japan for properly advancing resource management.

Furthermore, if this technique is developed further and data about the behavioral ecology and physiology of living marine resources can be measured on various spatial temporal scales, the relationships between marine environment changes and marine resource fluctuations will be surely understood more concretely. As fish form schools, evaluating the bio-logging data of an individual as a dynamic of a fish school and/or a population or measuring the fish school will be a crucial subject for future research.

Acknowledgments

I would like to thank Drs. H. Ueda and K. Tsukamoto for giving me the opportunity to write this chapter. I also thank the Fisheries Agency of Japan and the National Research Institute of Far Seas Fisheries, Japan, for allowing me to use the archival tag data. M. J. Miller, Atmosphere and Ocean Research Institute, The University of Tokyo, helped improve the manuscript. The author would like to thank Enago (www.enago.jp) for the English language review.

References

Aoki, I. and T. Murayama. 1993. Spawning pattern of the Japanese sardine *Sardinops melanostictus* off southern Kyushu and Shikoku, south western Japan. Mar. Ecol. Prog. Ser. 97: 127–134.

Barrett, I. and F.J. Hester. 1964. Body temperature of yellowfin and skipjack tunas in relation to sea surface temperature. Nature 203: 96–97.

Bayliff, W.H. 1994. A review of the biology and fisheries for northern bluefin tuna, *Thunnus thynnus*, in the Pacific Ocean. FAO Fish. Tech. Pap. 336: 244–295.

Bestley, S., T.A. Patterson, M.A. Hindell, and J.S. Gunn. 2008. Feeding ecology of wild migratory tunas revealed by archival tag records of visceral warming. J. Anim. Ecol. 77: 1223–1233.

Block, B.A., D.P. Costa, G.W. Boehlert, and R.E. Kochevar. 2003. Revealing pelagic habitat use: the tagging of Pacific pelagics program. Oceanologica Acta 25: 255–266.

Block, B.A. J.E. Keen, B. Castillo, H. Dewar, E.V. Freund, D.J. Marcinek, R.W. Brill, and C. Farwell. 1997. Environmental preferences of yellowfin tuna (*Thunnus albacares*) at the northern extent of its range. Mar. Biol. 130: 119–132.

Block, B.A., I.D. Jonsen, S.J. Jorgensen, A.J. Winship, S.A. Shaffer, S.J. Bograd, E.L. Hazen, D.G. Foley, G.A. Breed, A.L. Harrison, J.E. Ganong, A. Swithenbank, M. Castleton, H. Dewar, B.R.Mate, G.L. Shillinger, K.M. Schaefer, S.R. Benson, M.J. Weise, R.W. Henry, and D.P. Costa. 2011. Tracking apex marine predator movements in a dynamic ocean. Nature 475: 86–90.

Boustany, A. M. 2011. Bluefin Tuna: The State of the Science. Ocean Science Division, Pew Environment Group, Washington DC, USA.

Boustany, A.M., R. Matteson, M. Castleton, C. Farwell, and B.A. Block 2010. Movements of pacific bluefin tuna (*Thunnus orientalis*) in the Eastern North Pacific revealed with archival tags. Prog. Oceanogr. 86: 94–104.

Brill, R.W. 1994. A review of temperature and oxygen tolerance studies of tunas pertinent ot fisheries oceanography, movement models and stock assessments. Fish. Oceanogr. 3: 204–216.

Brill, R.W. 1996. Selective advantages conferred by the high performance physiology of tunas, billfishes, and dolphin fish. Comp. Biochem. Phys. A 113: 3–15.

Brill, R.W., H. Dewar, and J.B. Graham. 1994. Basic concepts relevant to heat-transfer in fishes, and their use in measuring the physiological thermoregulatory abilities of tunas. Environ. Biol. Fish. 40: 109–124.

Carey, F.G. 1973. Fishes with warm bodies. Sci. Am. 228: 36–44.

Carey, F.G., J.M. Teal, J.W. Kanwisher, and K.D. Lawson. 1971. Warm-bodied fish. Am. Zool. 11: 137–143.

Carey, F.G., J.W. Kanwisher, and E.D. Stevens. 1984. Bluefin tuna warm their viscera during digestion. J. Exp. Biol. 109: 1–20.

Clark, T.D., B.D. Taylor, R.S. Seymour, D. Ellis, J. Buchanan, Q.P. Fitzgibbon, and P.B. Frappell. 2008. Moving with the beat: heart rate and visceral temperature of free-swimming and feeding bluefin tuna. Proc. R. Soc. Lond. B. Biol. Sci. 275: 2841–2850.

Clark T.D., W.T. Brandt, J. Nogueira, L.E. Rodriguez, M. Price, C. J. Farwell, and B.A. Block 2010. Postprandial metabolism of Pacific bluefin tuna (*Thunnus orientalis*). J. Exp. Biol. 213: 2379–2385.

Dalton, R. 2005. Satellite tags give fresh angle on tuna quota. Nature 434: 1056–1057.

Domeier, M.L., D. Kiefer, N. Nasby-Lucas, A. Wagschal, and F. O'Brien. 2005. Tracking Pacific bluefin tuna (*Thunnus thynnus orientalis*) in the northeastern Pacific with an automated algorithm that estimates latitude by matching sea-surface-temperature data from satellites with temperature data from tags on fish. Fish. Bull. 103: 292–306.

Dragovich, A. 1969. Review of studies of tuna food in the Atlantic Ocean. U.S. Fish. Wildl. Serv. Spec. Sci. Rep. Fish. 593: 1–21.

Ekstrom, P.A. 2004. An advance in geolocation by light. Mem. Natl Inst. Polar Res. Spec. Issue. 58: 210–226.

Endo, Y. 2000. Distribution and standing stock: Japanese waters. *In*: Krill: Biology, Ecology and Fisheries. I. Everson [eds.]. Malden: Blackwell Science pp. 40–52.

Farwell, C.J. 2001. Tunas in captivity. *In*: Tuna: Physiology, Ecology, and Evolution. B.A. Block and E.D. Stevens [eds.]. San Diego, CA: Academic Press pp. 391–412.

Fishery Agency. 2009. Fisheries of Japan (FY2009), Fisheries policy for FY2010 (White Paper on Fisheries) 30 pp.

Gunn, J.S. and B. Block. 2001. Tuna metabolism and energetics. *In:* Tuna- Physiology, Ecology, and Evolution B.A. Block and E.D. Stevens [eds.]. San Diego: Academic Press pp. 167–224.

Goericke, R., S.J. Bograd, G. Gaxiola-Castro et al. 2004. The State of the California Current, 2003–2004: a rare "Normal" year. CalCOFI Rep. 45: 27–59.

Hamabe, M. 1964. Study on the migration of squid (*Ommastrephes slonani pacificus* Streernstrup) with reference to the age of the moon. Bull. Jpn Soc. Sci. Fish. 30: 209–215.

Hamasaki, S. and T. Nagai. 1995. Distribution and migration of the young bluefin tuna from the southwest area of Japan Sea to the mid-East China Sea. Bull. Jpn. Soc. Fish Oceanogr. 59: 398–408

Holland, K.N., R.W. Brill, R.K.C. Chang, J.R. Sibert, and D.A. Fournier. 1992. Physiological and behavioral thermoregulation in bigeye tuna (*Thunnus obesus*). Nature 358: 410–412.

Inagake, D., H. Yamada, K. Segawa, M. Okazaki, A. Nitta, and T. Itoh. 2001. Migration of young bluefin tuna, *Thunnus orientalis* Temminck et Schlegel, through archival tagging experiments and Its relation with oceanographic condition in the Western North Pacific. Bull. Natl. Inst. Far Seas Fish. 38: 53–81.

Itoh, T. 2001. Estimation of total catch in weight and catch at-age in number of bluefin tuna *Thunnus orientalis* in the whole Pacific Ocean. Bull. Natl. Inst. Far Seas Fish. 38: 83–111.

Itoh, T. 2006. Sizes of adult bluefin tuna *Thunnus orientalis* in different areas of the western Pacific Ocean. Fisheries Sci. 72: 53–62.

Itoh, T. 2009. Contributions of different spawning seasons to the stock of Pacific bluefin tuna *Thunnus orientalis* estimated from otolith daily increments and catch-at-length data of age-0 fish. Nippon Suisan Gakk. 75: 412–418.

Itoh, T., S. Tsuji, and A. Nitta. 2003a. Migration patterns of young Pacific bluefin tuna (*Thunnus orientalis*) determined with archival tags. Fish. Bull. 101: 514–534.

Itoh, T., S. Tsuji, and A. Nitta. 2003b. Swimming depth, ambient water temperature preference, and feeding frequency of young Pacific bluefin tuna (*Thunnus orientalis*) determined with archival tags. Fish. Bull. 101: 535–544.

Kawai, H. 1972. Hydrography of the Kuroshio extension. *In*: Kuroshio. H. Stommel and K. Yoshida [eds.]. Tokyo: University of Tokyo Press, Tokyo, Japan pp. 235–352.

Kinoshita, T. 1998. Northward migration juveniles in the Kuroshio Extension area. *In*: Stock fluctuations and ecological changes of the Japanese Sardine. Y. Watanabe and T. Wada [eds.]. Tokyo: Koseisya-Koseikaku pp. 84–92.

Kishinoue, K. 1923. Contributing to the comparative study of the so-called scombrid fishes. Journal of the College of Agriculture, Imperial University of Tokyo 8: 294–475.

Kitagawa, T. and S. Kimura. 2006. Alternative heat-budget model relevant to heat transfer in fishes and its practical use for detecting their physiological thermoregulation. Zool. Sci. 23: 1065–1071.

Kitagawa, T., H. Nakata, S. Kimura, T. Itoh, S. Tsuji, and A. Nitta. 2000. Effect of ambient temperature on the vertical distribution and movement of Pacific bluefin tuna *Thunnus thynnus orientalis*. Mar. Ecol. Prog. Ser. 206: 251–260.

Kitagawa, T., H. Nakata, S. Kimura, and S. Tsuji. 2001. Thermoconservation mechanisms inferred from peritoneal cavity temperature in free-swimming Pacific bluefin tuna *Thunnus thynnus orientalis*. Mar. Ecol. Prog. Ser. 220: 253–263.

Kitagawa, T., H. Nakata, S. Kimura, T. Sugimoto, and H. Yamada. 2002. Differences in vertical distribution and movement of Pacific bluefin tuna (*Thunnus thynnus orientalis*) among areas: the East China Sea, the Sea of Japan and the western North Pacific. Mar. Freshwater Res. 53: 245–252.

Kitagawa, T., S. Kimura, H. Nakata, and Yamada. 2004a. Overview of the research on tuna thermo-physiology using electric tags. Mem. Natl Inst. Polar Res. Spec. Issue. 58: 69–79.

Kitagawa, T., S. Kimura, H. Nakata, and Yamada. 2004b. Diving behavior of immature, feeding Pacific bluefin tuna (*Thunnus thynnus orientalis*) in relation to season and area: the East China Sea and the Kuroshio-Oyashio transition region. Fish. Oceanogr. 13: 161–180.

Kitagawa, T., S. Kimura, H. Nakata, and H. Yamada. 2006a. Thermal adaptation of Pacific bluefin tuna *Thunnus orientalis* to temperate waters. Fish. Sci. 72: 149–156.

Kitagawa, T., A. Sartimbul, H. Nakata, S. Kimura, H. Yamada, and A. Nitta. 2006b. The effect of water temperature on habitat use of young Pacific bluefin tuna *Thunnus orientalis* in the East China Sea. Fish. Sci. 72: 1166–1176

Kitagawa, T., A.M. Boustany, C.J. Farwell, T.D. Williams, M.R. Castleton, and B.A. Block. 2007a. Horizontal and vertical movements of juvenile bluefin tuna (*Thunnus orientalis*) in relation to seasons and oceanographic conditions in the eastern Pacific Ocean. Fish. Oceanogr. 16: 409–421.

Kitagawa, T., S. Kimura, H. Nakata, and H. Yamada. 2007b. Why do young Pacific bluefin tuna repeatedly dive to depths through the thermocline? Fish. Sci. 73: 98–106.

Kitagawa, T., S. Kimura, H. Nakata, H. Yamada, A. Nitta, Y. Sasai, and H. Sasaki. 2009. Immature Pacific bluefin tuna, *Thunnus orientalis*, utilizes cold waters in the Subarctic Frontal Zone for trans-Pacific migration. Environ. Biol. Fish. 84: 193–196.

Kitagawa, Y., Y. Nishikawa, T. Kubota, and M. Okiyama. 1995. Distribution of ichthyoplanktons in the Japan Sea during summer, 1984, with special reference to scombroid fishes. Bull. Jpn. Soc. Fish Oceanogr. 59: 107–114.

Koido, T. and K. Mizuno. 1989. Fluctuation of catch for bluefin tuna (*Thunnus thynnus*) by trap nets in Sanriku coast with reference to hydrographic condition. Bull. Jpn. Soc. Fish Oceanogr. 53: 138–152.

Korsmeyer, K.E. and H. Dewar. 2001. Tuna metabolism and energetics. *In*: Tuna-physiology, ecology, and evolution. B.A. Block and E.D. Stevens [eds.]. San Diego: Academic Press pp. 35–78.

Majkowski, J. 2007. Global fishery resources of tuna and tuna-like species. FAO Fish. Tech. Pap. 483

Marcinek, D.J., S.B. Blackwell, H. Dewar, E.V. Freund, C. Farwell, D. Dau, A.C. Seitz, and B.A. Block. 2001. Depth and muscle temperature of Pacific bluefin tuna examined with acoustic and pop-up satellite archival tags. Mar. Biol. 138: 869–885.

Matsumura, Y. 1989. Factor affecting catch of young tuna *Thunnus orientalis* in waters around the Tsushima Islands. Nippon Suisan Gakk. 55: 1703–1706.

The Ministry of Agriculture, Forestry and Fisheries. 1997–2001. Preliminary Statistical Report on the Production in Fisheries and Aquaculture 1995–1999. Preliminary Statistical Report on the Production in Fisheries and Aquaculture. Tokyo: The Ministry of Agriculture, Forestry and Fisheries.

Miyake, M., N. Miyabe, and H. Nakano. 2004. Historical trends of tuna catches in the world. FAO Tech. Pap. 467.

Musyl, M.K., R.W. Brill, D.S. Curran, J.S. Gunn, J.R. Hartog, R.D. Hill, D.W. Welch, J.P. Eveson, C.H. Boggs, and R.E. Brainard. 2001. Ability of archival tags to provide estimates of geographical position based on light intensity. *In*: Electronic tagging and tracking in marine fisheries reviews: methods and technologies in fish biology and fisheries. J.R. Sibert and J.L. Nielsen [eds.]. Dordrecht: Kluwer Academic Press pp. 343–368.

Nishimura, S. 1965. The zoogeographical aspects of the Japan Sea. Part I. Publ. Seto Mar. Biol. Lab 13: 35–79.

Nozu, J. 1966. On the swimming layer of Engraulis Japonica shoals in Bungo Channel and its adjacent waters. Bull. Jpn. Soc. Sci. Fish. 32: 237–241.

Ogawa, Y. and T. Ishida. 1989a. Distinctive features of fluctuations in the catch of *Thunnus thynnus* by set-nets along the Sanriku coast. Bull. Tohoku Reg. Fish. Res. Lab. 51: 11–21.

Ogawa, Y. and T. Ishida. 1989b. Hydrographic conditions governing fluctuations in the catch of *Thunnus thynnus* by set-nets along the Sanriku coast. Bull. Tohoku Reg. Fish. Res. Lab. 51: 22–39.

Ohshimo, S. 1996. Acoustic estimation of biomass and school character of anchovy Engraulis japonicus in the East China Sea and the Yellow Sea. Fish. Sci. 62: 344–349.

Ohshimo, S. 1998. Distribution and stomach contents of *Maurolicus muelleri* in the Japan Sea. J. Korean Soc. Fish. Res. 1: 168–175.

Okiyama, M. 1974. Occurrence of the postlarvae of bluefin tuna, *Thunnus thynnus*, in the Japan Sea. Bull Jpn. Sea. Reg. Fish Res. Lab. 11: 9–21.

Polovina, J.J. 1996. Decadal variation in the trans Pacific migration of northern bluefin tuna (*Thunnus thynnus*) coherent with climate induced change in prey abundance. Fish Oceanogr 5: 114–119

Smith, P.J., A.M. Conroy, and P.R. Taylor. 1994. Biochemicalgenetic identification of northern bluefin tuna *Thunnus thynnus* in the New Zealand fishery. New Zealand J. Mar. Freshwater Res. 28: 113–118.

Shimose, T., T. Tanabe, K. S. Chen, and C.C. Hsu. 2009. Age determination and growth of Pacific bluefin tuna, *Thunnus orientalis*, off Japan and Taiwan. Fish. Res. 100: 134–139.

Stevens, E.D., H.M. Lam, and J. Kendall. 1974. Vascular Anatomy of Countercurrent Heat-Exchanger of Skipjack Tuna. J. Exp. Biol. 61: 145–153.

Stevens, E.D., J.W. Kanwisher, and F.G. Carey. 2000. Muscle temperature in free-swimming giant Atlantic bluefin tuna (*Thunnus thynnus* L.). J. Therm. Biol. 25: 419–423.

Sund, P.N., M. Blackburn, and F. Williams. 1981. Tunas and their environment in the Pacific Ocean: a review. Oceanogr. Mar. Biol. An. Ann. Rev. 19: 443–512.

Suzuki, T., M. Tashiro, and Y. Yamagishi. 1974. Studies on the swimming layer of Squid *Todarodes pacificus* Steenstrup as observed by a fish finder in the offshore region of the northern part of the Japan Sea. Bull. Fac. Fish. Hokkaido Univ. 25: 238–246.

Takahashi, M., Y. Watanabe, T. Kinoshita, and C. Watanabe 2001. Growth of larval and early juvenile Japanese anchovy, *Engraulis japonicus*, in the Kuroshio-Oyashio transition region. Fish. Oceanogr. 10: 235–247.

Tanaka, Y., K. Satoh, M. Iwahashi, and H. Yamada. 2006. Growth-dependent recruitment of Pacific bluefin tuna *Thunnus orientalis* in the northwestern Pacific Ocean. Mar. Ecol. Prog. Ser. 319: 225–235.

Teo, S.L.H., A. Boustany, S. Blackwell, A. Walli, K.C. Weng, and B.A. Block. 2004. Validation of geolocation estimates based on light level and sea surface temperature from electronic tags. Mar. Ecol. Prog. Ser. 283: 81–98.

Teo, S.L.H., A. Boustany, H. Dewar, M.J.W. Stokesbury, K.C. Weng, S. Beemer, A.C. Seitz, C.J. Farwell, E.D. Prince, and B.A. Block. 2007. Annual migrations, diving behavior, and thermal biology of Atlantic bluefin tuna, *Thunnus thynnus*, on their Gulf of Mexico breeding grounds. Mar. Biol. 151: 1–18.

Tsuji, S. and T. Itoh. 1998. Ecology and recruitment fluctuation of northern bluefin tuna. *In:* Seikai National Fisheries Research Institute, Fishery Agency of Japan [eds.]. Proceedings of Japan–China Joint Symposium on CSSCS, Nagasaki, Japan pp. 321–330.

Uda, M. 1957. A consideration on the long years trend of the fisheries fluctuation in relation to sea condition. Bull. Jpn. Soc. Sci. Fish. 23: 68–72.

Uda, M. 1973. Pulsative fluctuation of oceanic fronts in association with the tuna fishing grounds and Fisheries. J. Fac. Mar. Sci. Tech. Tokai Univ. 7: 245–267.

Ueyanagi, S. 1969. Observations on the distribution of tuna larvae in the Indo-Pacific Ocean with emphasis on the delineation of the spawning areas of albacore, *Thunnus alalunga*. Bull. Far. Seas. Fish. Res. Lab. 2: 177–256.

Welch, D.W. and J.P. Eveson. 1999. An assessment of light-based geoposition estimates from archival tags. Can. J. Fish. Aquat. Sci. 56: 1317–1327.

Yabe, H., S. Ueyanai, and H. Watanabe. 1966. Studies on the early life history of bluefin tuna *Thunnus thynnus* and on the larvae of the southern bluefin tuna *T. maccoyii*. Rep. Nankai. Reg. Fish. Res. Lab. 23: 95–129.

Yamada, H., N. Takagi, and D. Nishimura. 2006. Recruitment abundance index of Pacific bluefin tuna using fisheries data on juveniles. Fish. Sci. 72: 333–341.

Yamanaka, H. 1982. Fishery biology of the bluefin tuna resource in the Pacific Ocean. Japan Fisheries Resources Conservation Association, Tokyo, Japan.

Yamanaka, H. et al. 1963. Synopsis of biological data on kuromaguro *Thunnus orientalis* (Temminck & Schlegel) 1942 (Pacific Ocean). FAO Fish. Rep. 6: 180–217.

Yasui, M., T. Yasuoka, K. Tanioka, and O. Shiota. 1967. Oceanographic studies of the Japan Sea (1). Oceanograpical Magazine 18: 177–192.

Yokota, T., M. Toriyama, F. Kanai, and S. Nomura. 1961. Studies on the feeding habit of fishes. Rep. Nankai Reg. Fish. Res. Lab. 14: 1–23.

Yukinawa, M. and Y. Yabuta 1967. Age and growth of bluefin tuna, *Thunnus thynnus* (Linnaeus), in the north Pacific Ocean. Rep. Nankai Reg. Fish. Res. Lab. 25: 1–18.

Zar, J.H. 1996. Biostatistical analysis, 3rd edn. Englewood Cliffs, NJ: Prentice-Hall, 662 pp.

Index

Color Plate Section

Chapter 2

Fig. 3. The River Tees Barrage in north east England, an amenity barrage that provides challenges to the upstream migration of spawning Atlantic salmon. During periods of low flow when fish congregate below the barrage, predation by seals is common.

Chapter 3

Fig. 3. Photograph of different stages of the yellow (top) and silver (bottom) Japanese eels, *Anguilla japonica* (Okamura et al. 2007).

Chapter 4

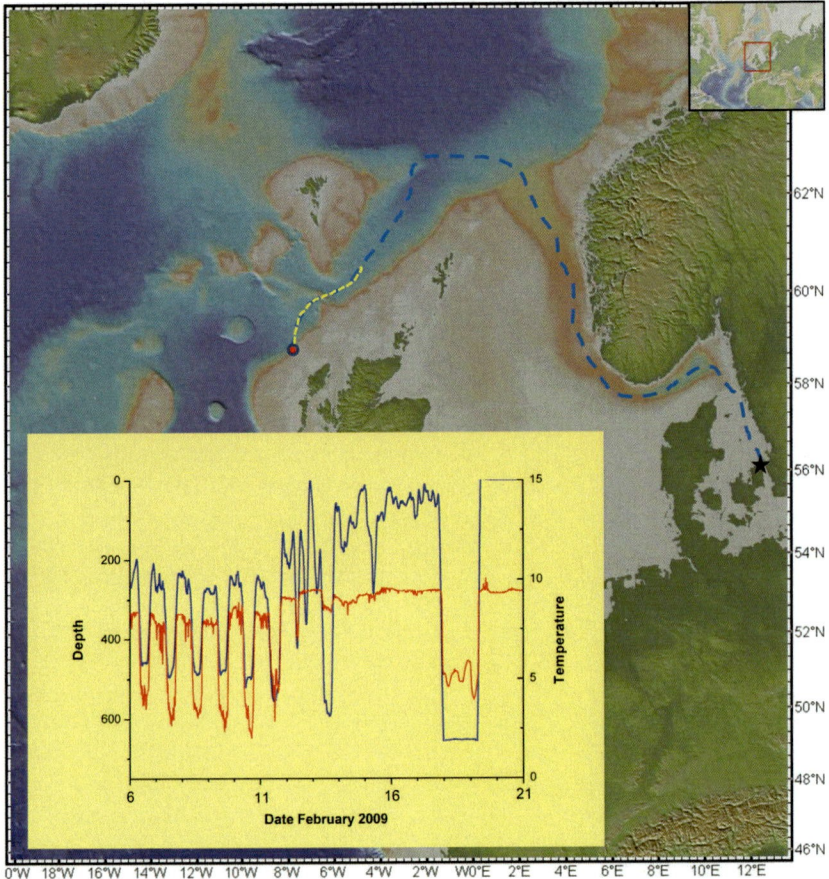

Fig. 6. The reconstructed trajectory of a silver eel tagged internally with a fl oating DST at the outlet of the Baltic (star). The position where the tag surfaced is shown as a red point. The inserted swimming depth and temperature diagram shows the portion of the track highlighted in yellow. For discussion see text.